A Flood Cannot Happen Here

The Story of Lower Goose Creek Reservoir, Oakley, Idaho, 1984

Kathleen Hedberg

Magic Valley Publishers
BURLEY, IDAHO

99 98 97 96 95 94 93 10 9 8 7 6 5 4 3 2 1

Cover photo: Goose Creek above the reservoir, spring 1984.
(Courtesy of Ryan Thomas.)

Photo on page 280: West canal at the diversion point.
(Courtesy of Donald J. Clark.)

Cover design: Teddie Kraus

Maps: Kent Schmidt

Drawing: Marilyn R Miller

Graphs: Shane Olsen

Typesetting and editing: Leon Johnson, Laser Graphics

Printing: Publishers Press

Magic Valley Publishers
P.O. Box 501
Burley, Idaho 83318
Phone: 208 678-2406

Library of Congress number: 93–077900

ISBN 0–9636310–0–4
ISBN 0–9636310–1–2 (paperback)

Contents

Preface v
Introduction 1
1. Fall 1983 5
2. Early Days 9
3. Winter, 1983–1984 25
4. April 1984 37
5. May 1–May 6 47
6. May 7–May 12 73
7. May 13–May 15 103
8. Wednesday, May 16 129
9. Thursday, May 17 151
10. Friday & Saturday, May 18–19 177
11. Sunday, May 20 199
12. May 21–June 8 215
13. June–December 251
Epilogue 279
Appendix A 283
Appendix B 285
Appendix C 291
Sources Consulted 293

Preface

This book is different from the book I intended to write. I planned to write an adventure story, a true-life story of man against nature, an uncomplicated story about an event. I thought it would be easy. All I had to do was interview about fifty people, read some newspaper accounts, collect a few pictures, and put it together.

I soon found it would not be so simple. Instead of one big story, there were hundreds of little stories, all of which were needed to form the whole. Of greater importance than the events that occurred were the people whose actions and reactions shaped those events. Choosing what to include was difficult. So much happened; so many people were involved. Yet so much had to be left out, so many people and their contributions unmentioned and unrecognized. The ones spotlighted in this account are examples of a much broader group; their stories are a microcosm of the whole.

The events of 1984 touched people deeply, changing their lives in significant ways. Writing about these events has changed my life also, making me more aware of the goodness of people, more appreciative of blessings from the Lord.

My thanks go to those who have helped in the preparation and publication of this book. Leon Johnson of Laser Graphics has been more than an editor and typesetter; he has been a friend who advised and encouraged me and did hundreds of little things to make this book the best it could be. Teddie Kraus designed the cover and taught me how to select and crop pictures. Leonard Arrington and Robert Woody read the manuscript and as "outsiders" gave me suggestions; Richard Hedberg advised me on business procedures and finance; Joan Matthews and Eileen Petersen helped with editing and proofreading. Many people have loaned pictures, shared their stories, and given me technical help, advice, and encouragement. I express my thanks to all.

But most especially I am thankful to the people who were part of the events of 1984. To the people of our valley and to the others who helped during our time of need, this book is gratefully dedicated.

Introduction

A Flood Cannot Happen Here is the story of a community that differs in many ways from other parts of the United States. Although its people own computers and worry about drugs, their lifestyle and values are more like those of an earlier time.

It is a rural community. With 2,500 square miles, Cassia (ca′ zha) County is larger than the state of Delaware, but its 20,000 population equals .03 percent of Delaware's population. Cassia County and neighboring Minidoka (min a do′ ka) County form a distinct geographic unit.

About 18 percent of the people in Cassia County are farmers, compared with 2 percent in the United States as a whole. Most Cassia County farmers have relatively small holdings; they are family farmers who live on their land. Unlike the farmers of the Midwest, they depend almost entirely on irrigation.

Farming and farm-related businesses form the economic base of Cassia County; when the farmers prosper, everyone prospers. The industrial recession of the early 1990s has hardly been felt in this community, but the farm-based recession of the early 1980s hit the area hard. Farmers and others were still feeling the pinch of hard times in 1984 when the events occurred that are related in this book.

Religion plays an important role in people's lives. Although there are sizeable congregations of Catholic and most Protestant denominations, The Church of Jesus Christ of Latter-day Saints (LDS or Mormon) is the largest religious group. With about 3,400 members in Oakley and the rural area, Mormons comprise about 86 percent of the population. With about 4,600 members in Burley, they make up about 50 percent of Burley's population.

The religious and social lives of LDS church members revolve around their local congregation, which is called a ward. The names of some early wards—Star, Pella, Unity, and View in western Cassia County—are also the names of farming areas. The wards, which average 300 to 600 members, provide Sunday worship services and weekly youth activities. Men as "home teachers" and women as "visiting teachers" visit all families in the ward every month, giving them spiritual messages and checking on their welfare. All active male members of the church hold the Priesthood which includes the power to bless, to perform ordinances, and to lead church congregations. The men are organized in Priesthood quorums of High Priests, who are usually older men, and Elders, who are usually younger men; the women form the Relief Society.

At the head of each ward is a bishop who is assisted by two counselors; five to twelve wards comprise a stake which is headed by a stake president with two counselors. General authorities in Salt Lake City, Utah, appoint these local leaders who usually serve for three to ten years. Because the local leaders are not paid by the church, each continues to work in his full-time occupation.

In most other religious denominations in Cassia County, professional clergy head the congregations. Their members, however, also engage in voluntary church service, sponsor youth activities, and have a social as well as religious cohesiveness.

In addition there are forty-one service organizations in the Mini-Cassia area as well as numerous literary, social, and special interest clubs. These groups cut across denominational lines, uniting people of all religious beliefs.

In an age when traditional ties are fragmenting, and at a time when many people have a greater loyalty to those of the same occupational, ethnic, or cultural background than to their neighbors, the people of the Mini-Cassia area maintain a loyalty to place. Their ties to their neighbors are stronger than their religious, economic, or social differences. In an age of social fragmentation, they have retained a sense of community. And it was as a community that they faced the challenge of flooding in 1984.

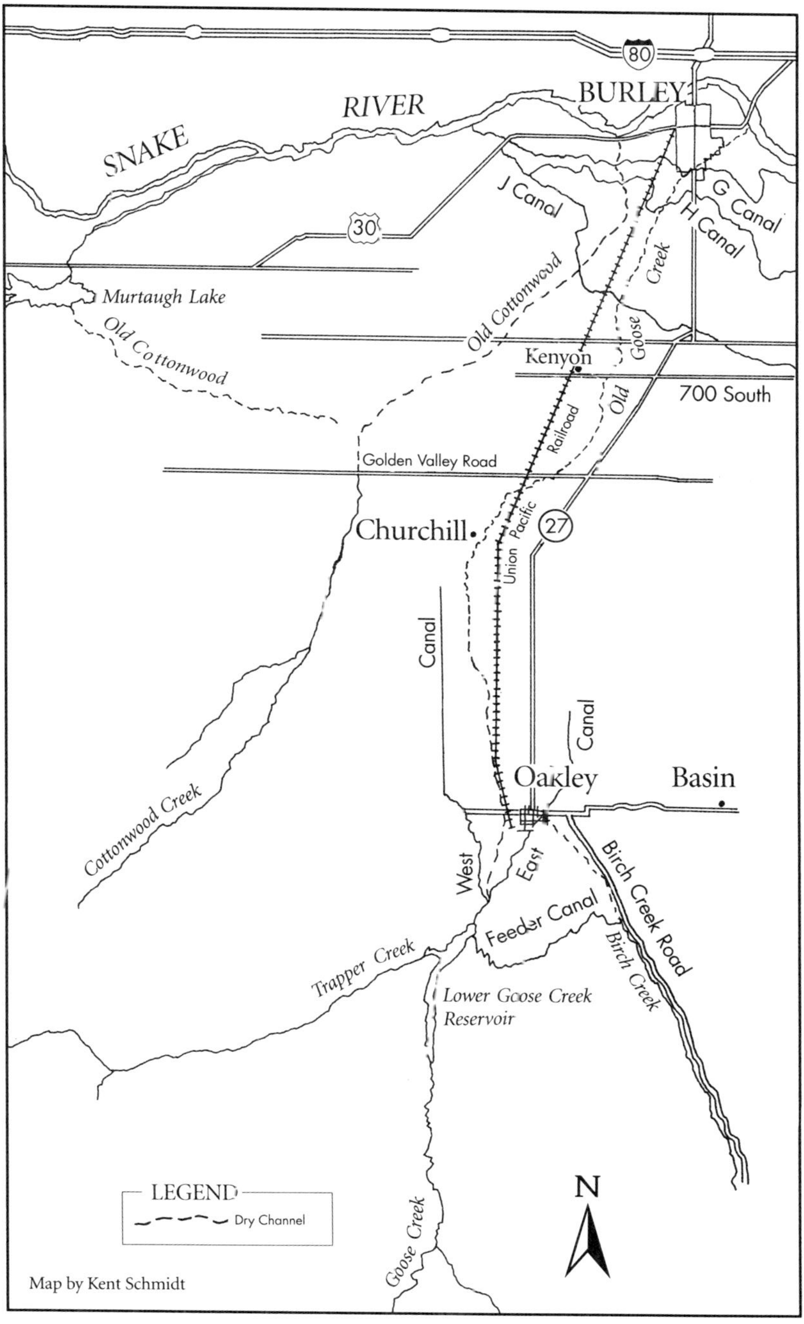
80
BURLEY
SNAKE
RIVER
30
J Canal
G Canal
H Canal
Murtaugh Lake
Old Cottonwood
Old Cottonwood
Creek
Goose
Old
Kenyon
700 South
Railroad
Golden Valley Road
Union Pacific
Churchill
27
Canal
Canal
Oakley
Basin
Cottonwood Creek
West
East
Birch Creek Road
Feeder Canal
Birch Creek
Trapper Creek
Lower Goose Creek Reservoir
LEGEND
Dry Channel
Goose Creek
N
Map by Kent Schmidt

Lower Goose Creek Reservoir, spring 1984.
(Photo courtesy of Nona Jenks.)

Chapter One

Fall 1983

No one in Cassia County was worried about a flood. Major floods did not happen in the dry lands of south-central Idaho. When 2.23 inches of rain, 50 percent more than usual, fell in September and October of 1983, everyone was happy. Cassia County residents prayed regularly for rain and snow. It did not occur to them that there could be too much.

The early storms pleased Deputy Sheriff Terry Bingham. As a young college graduate Terry had moved to Cassia County in 1974 to work in law enforcement. As volunteer assistant civil defense director for Cassia County, Terry also planned for disasters such as flood, earthquake, and nuclear contamination.

Of all possible disasters, a major flood seemed the least likely. Yet a flood had occurred in eastern Idaho, in an area that also had seemed safe. In 1976 the newly-constructed Teton Dam failed, sending flood waters through farms and homes, killing eleven people, and severely damaging the city of Rexburg.

Many people from Cassia County had helped Rexburg residents haul mud from basements, cart away debris, scrub walls and furniture. They remembered the smashed cars and houses, the smell, and the vast grayness that had covered everything.

"Travel homes were mashed together like aluminum cans," said Burley resident David Gibbons. "One car did not look like it had been hurt. I opened the door, and the mud was level with the seats. One home had not been washed away because a fence corner diverted the flood waters. A little boy had been playing with some toy trucks in the front yard. The trucks were still there, untouched. But the family car, parked a few feet away, had been caught by the water and washed down the road for miles."

Several people stayed in Rexburg for weeks to help with the cleanup. "People in our neighborhood took up equipment," said Pella farmer Fred Hawker. "I didn't have equipment, but I took up my motor home so they'd have a place to sleep. We'd take turns—go up for awhile, come back and farm for awhile, then go up again."

Others went for a day at a time in groups organized by the LDS church. Hundreds of school buses brought volunteers from communities as far as two hundred miles away. On her first trip one woman watched all the buses converge on Rexburg and wondered how there could possibly be enough work for that many people. After spending most of the day cleaning hard, sticky mud off one antique rocker, she wondered how any number of people could do everything that needed to be done. Where she worked, the house had been filled with about five feet of water. Above the water line, the rooms looked normal, with pictures and painted walls. Below the line everything was covered with gray, sticky mud.

"We took three or four bus loads and helped dig the town of Roberts out," said the Reverend Leman Messley, at that time minister of the Christian Church (Disciples of Christ). He remembered the mud being so slimy, the houses oozing mud. "But my most vivid memory was the depression of the people," he said. "This was right after it happened. They appeared to be in shock, completely devastated by the thing. Nobody could tell us what to do. We pulled the pump off the city well, started repairing their water system. As the day went by, people began to think, 'Maybe this is do-able after all.'"

"Some of the houses didn't seem to be hurt too bad," said Leland Baker, Mini-Cassia Search and Rescue volunteer. "At first we thought all we had to do was tear out the carpet. Then we'd find that the wall board, plaster board, and sheet rock was all ruined. The insulation was soaked and falling off, electrical and gas lines destroyed. It was not as minor as it appeared when we first surveyed the damage."

"In my way of thinking, a flooded basement is one notch worse than a house fire," said Larry Adams. "With a fire you know you've got to start all over again, but you try to save things in a flooded basement."

D. Rex Gerratt (Jer′ et), president of the Oakley Stake of the LDS church,was not thinking about either fire or flood during the fall of 1983. Born in Minidoka County in 1936, Rex had moved with his family to the Unity area south of Burley when he was eleven. He and his brother Larry owned a dairy, farming, and livestock-raising operation.

Rex had recently been appointed bi-regional emergency director for the Burley and Rupert regions of the LDS church. He had no concern about his new calling. "It sounded to me as if I got the least difficult of the regional assignments," he later said. When he tried to think of a possible future catastrophe in the region, he could not think of one. "My thoughts went back to Rexburg, but I knew we were not in any danger here. A flood could not happen in our area."

J. Weldon Beck was not thinking about a flood either. As one of three Cassia County commissioners, he worried instead about the economy. Idaho's economy had been depressed when the 1981–82 recession pushed the state into its worst downturn in recent history. Although by 1983 the recession was over for the industrial sections of the nation, the recession continued in Idaho.

Weldon had seen hard times before. He had graduated from college during the Great Depression, and had taught school until he purchased a small farm. He had been elected county commissioner in 1959, and had served continually since, having run unopposed in most elections.

Many changes had occurred during his twenty-three years in office. In earlier times people took care of themselves, so the commissioners had fewer problems with indigence and welfare. Lately the biggest problem had been liability. It seemed to Weldon that people were ready to sue the county and the commissioners over any little thing.

Weldon was pleased, that fall of 1983, as he watched the snow come down. With so much early snow, no one would have to worry about water this year.

A country road near Oakley, 1909.
(Photo courtesy of the Oakley Canal Company.)

Chapter Two

Early Days

Water brought the first settlers to Oakley, water in the mountain streams that flowed into the valley. The largest of these was Goose Creek, which originated in high mountains on the Utah, Idaho, and Nevada borders. Goose Creek flowed north through mountains and rangelands, through a lava-rock canyon above Oakley into the flat, fan-shaped Oakley valley, and north through the Snake River plain. Through thousands of years of geologic time, Goose Creek had meandered through this area, depositing its rich, black soil, building up and filling one channel, cutting another, taking forty-three miles to go the twenty-three miles from its canyon above Oakley to the Snake River.

James Dayley and sons were among the first settlers in the Oakley valley. Arriving in 1879, they sent word to relatives in Utah that they had found a good place. Others came the next year; 150 people were living in the valley by 1881. Most were Mormons, members of The Church of Jesus Christ of Latter-day Saints. In 1881 church leaders asked Horton D. Haight to move to Oakley to become bishop and later stake president for the growing community.

The early settlers raised alfalfa to feed their livestock and gardens to feed themselves. They irrigated their fields from Goose Creek and other streams. Water was always uncertain. Melting snow produced heavy runoff in the spring, but creeks slowed to a trickle or disappeared during the hot summer months. The settlers stored water in cisterns and behind small dams. "The water is the only drawback we are afflicted with," the *Oakley Star* noted in 1893.

The settlers prospered. When a coal mine, a marble quarry, and a precious metals mine opened in the vicinity, Oakley became the trading center for the region. By 1905 Oakley had two thousand people.

Meanwhile a new town was rising in the northern end of the valley on the delta where Goose Creek entered the Snake River. Burley, as the town was called, would soon have something that Oakley did not have—a dependable water supply for irrigation.

In 1904, the federal government had begun the construction of Minidoka Dam to furnish irrigation water for approximately 120,000 acres on both sides of the Snake River. Officials platted the Burley townsite in 1905 as homesteaders from all over the nation filed for land. With the completion of Minidoka Dam in 1906, homesteaders began farming on the north side of the river in what would become Minidoka County. But because the land was higher on the south side of the river in Cassia County, pumps were needed to lift water to that area. The Minidoka powerplant supplied electricity to lift the water to canals thirty, sixty, and ninety feet above the river. Although the irrigation officials named them the G, H, and J canals, farmers called them the first-, second-, and third-lift canals. These canals brought irrigation water to the Burley area in 1910.

Under provisions of the Carey Act, private companies could also develop water projects. In 1908 William and James Kuhn, owners of the American Water Works and Guarantee Company of Pittsburgh, Pennsylvania, formed the Twin Falls-Oakley Land and Water Company to construct a dam seven miles southwest of Oakley on Goose Creek.* In 1909 the state of Idaho segregated 43,893 acres for the project. Included in this segregation were 10,000 acres which the Oakley settlers owned.

---

* The Twin Falls South Side Project, the first major Carey Act project in Idaho, was a great success. Even though the Oakley project had no connection with Twin Falls, the Kuhns used "Twin Falls" as part of their company name to take advantage of favorable publicity.

The Twin Falls-Oakley Land and Water Company purchased the settlers' Goose Creek water rights, giving them company stock which guaranteed delivery of 1½ acre feet of water per share. Most Oakley farmers raised grain and alfalfa and grazed sheep and cattle on the open range; 1½ acre feet of water per year seemed adequate for these crops. (One acre foot of water is the amount of water it takes to cover one acre of land with water one foot deep.)*

Oakley people were enthusiastic about the project, remembered John Clark, who was ten years old when construction on the dam started. "They anticipated that water would cover the whole valley. A few people resisted giving up their water rights—one man never did sell. But 98 to 99 percent of the people went for it." The *Oakley Herald* predicted that within ten years Oakley would have 10,000 to 30,000 inhabitants.

Investors in the East and Midwest also greeted the project with enthusiasm. Company agents sold bonds for the $1.5 to $2 million the project would cost and convinced prospective settlers to buy stock for water shares. The first issue of stock sold for $40 a share; in 1913 stock sold for $65 a share. Three new settlements—Churchill, Golden Valley, and Kenyon, with stores, schools, and post offices—were established north of Oakley.

Dam construction began in 1910. It was to be an earth-fill dam 1,050 feet long, 733 feet thick at its base, 143 feet high, with a

---

* Several terms indicate volume and flow of water. "Acre feet" is a measure of volume, and is generally used to describe the amount of water in a reservoir. One acre foot of water is 43,560 cubic feet, or 325,900 gallons. The Oakley Canal Company measures water in the reservoir in acre feet, and calculates water shares in acre feet, or in "acre inches." One share of water is 1½ acre feet, or 18 inches.

Government officials and scientists generally use "cubic feet per second (cfs)" to describe water flow. One cfs indicates that one cubic foot of water, or 7.48 gallons, is flowing past a given point in one second. One cfs flowing for twenty-four hours equals 1.983471 acre feet or approximately 2 acre feet; 100 cfs flowing for twenty-four hours equals approximately 200 acre feet of water.

Because farmers work with small streams of water, they use the term "miners inch" to describe flowing water. A miners inch is an old measurement which is not the same in every state. In Idaho 50 miners inches equals 1 cfs. Thus 5,000 miners inches equals 100 cfs, or 200 acre feet of water in a twenty-four-hour period.

Construction of the Lower Goose Creek Reservoir and canal system, 1910. Top: Hauling concrete from the mixer to the corewall. Center: Diverting weir and headgates for the High Line Canal. Bottom: Backfilling along the corewall. (Photos courtesy of the Oakley Canal Company.)

1-foot-thick concrete core. It could contain 74,500 acre feet of water.

Water for irrigation was supposed to flow through a tower when the water level was high and through gates near the bottom of the reservoir when the water level was low. The tower was used only once. "Air in the tower caused the water to surge like it was coming out of a Coke bottle turned upside down," said watermaster Clem Stanger. "It shook the tower and the machinery—scared the operators. From then on they used the gates to regulate the water even when the reservoir was high."

Surplus water was supposed to flow over a 600-foot-long spillway which was built along a ravine on the west side of the reservoir. The developers were sure there would be surplus water. They named the reservoir Lower Goose Creek Reservoir, confident that in a few years an upper Goose Creek reservoir would be needed. They expected that the runoff into the reservoir from four creeks—Birch, Cottonwood, Goose, and Trapper—would average 148,000 acre feet.*

When the dam was half completed, John Clark's parents took the family to see it. John remembered watching the horse and mule teams—one driven by his older brother Saul. He saw water wagons sprinkling dirt and an electric-powered conveyor belt bringing fill material from a hill about a quarter mile away. He heard people say that it would be the largest earthen dam in the world.

---

* In a thesis published in 1913, project engineer Amund Korsmo described how the developers calculated the water supply. Oakley weather records from 1895-1909 showed that Oakley averaged 9.22 inches of precipitation per year. A Bureau of Reclamation official had demonstrated that in the Sierra Nevadas in California, precipitation increased .6 inches for every 100-foot rise in elevation. Because the average elevation of the Goose Creek drainage area was 1,800 feet higher than Oakley, engineers calculated that precipitation would average 20 inches per year. Using another formula, they estimated that runoff would average 148,420 acre feet. They expected a runoff of 105,000 acre feet from Goose Creek, 30,000 acre feet from Cottonwood and Trapper creeks combined, and 13,420 acre feet from Birch Creek. If one-fourth of the water soaked into the ground, 113,537 acre feet of water would be available for farming, enough to provide water for 75,700 acres. The developers thought they were being conservative when they requested a segregation of only 50,000 acres.

View of Lower Goose Creek Reservoir from upstream, 1911. (Photo courtesy of the Oakley Canal Company.)

By 1912 the dam was complete enough to store water fifty-six feet deep. "From the viewpoint of having plenty of water there never was a time when the prospects for a splendid harvest was brighter than this year," reported the *Oakley Herald*. "It is quite evident that there is plenty of water to irrigate all the land between Oakley and the third lift on the Minidoka project."

"Fifty Thousand Acres of Rich Land in Cassia County to be Reclaimed by Building of Wonderful Dam and Magnificent Irrigation System" proclaimed the *Burley Bulletin* in March 1913. "The waters from hundreds of square miles of the highest mountains in southern Idaho and northern Nevada are drained into the reservoir, assuring an ample supply even during the driest seasons."

Three months later project developer William Kuhn, who had financial problems with other water projects, filed for bankruptcy. Because the Oakley project was almost complete, the bondholders formed an association to furnish the money needed to finish the irrigation system.

The reservoir filled partially in 1913. "I was only a baby when the dam was built," said Martha Rose Lattimer, "but I heard Mother say that they put all the children in the buggy, and took them to the reservoir to watch the water come in. It was a big deal. They thought the water would be so plentiful. To see it all was so wonderful." Her mother wanted the children to remember the historic occasion when Oakley would have enough water.

But the historic occasion did not arrive; the reservoir did not fill. Only a little Birch Creek water flowed into the reservoir. Farmers on Birch Creek, whose lands were higher than the reservoir, had retained their water rights. Water from Cottonwood Creek flowed into the feeder canal, but disappeared into the ground before it reached the reservoir. Water from Goose and Trapper creeks flowed into the reservoir as anticipated, but those creeks did not have as much volume as the engineers had projected. In 1913 combined runoff from Trapper and Goose creeks was 50,377 acre feet. It was 65,889 acre feet in 1914; 28,532 acre feet in 1915—far short of the 148,000 acre feet that had been projected. Between 1915 and 1920 farmers received full water rights only once.

"It was a great disappointment to the people when they realized there would not be enough water," said Eula Matthews Clark, who was a child at the time. "But they kept looking forward to next year, thinking it would be better."

The "old settlers" in Oakley could manage. Their homes and farms were established; they could cope with inadequate water. But the "new settlers" were working to clear land and build homes, a difficult task under the best of conditions. The scarcity of water in addition to the difficulties of clearing land was too much. They could not pay their water assessments.

The bondholders knew they could get their investment money back only if the settlers paid. The settlers argued that they should not have to pay unless they received full water rights.

Over the next several years lawyers and state investigators tried to solve the problem of too little water for too much land.* The old settlers wanted the project reduced to 15,000 acres. The bond-

---

* For a detailed description of these legal disputes, see Hugh T. Lovin, "A 'New West' Reclamation Tragedy: The Twin Falls-Oakley Project in Idaho 1908-1931." *Arizona and the West* 20 (Spring 1978): 5-24. This article is based mainly on court documents. From his study of these, Lovin concludes that the Oakley settlers did not want their lands to be a part of the reclamation project. However, the children of Oakley settlers remember their parents telling them how thrilled most people were at the prospects of having a dam built, and that only a few were opposed. Likely the extent of the opposition was exaggerated in court when it became evident that the reservoir would not fill.

holders, whose income was based on the project's size, wanted it to remain at the 28,000 acres that had not been abandoned. Eventually water was taken off the lands at Kenyon and the project reduced to 21,000 acres. Those who lost their water rights received only the fifty cents per acre that they had originally paid for the land.

Throughout most of the nation, the 1910s were good years for farmers. But in 1920 commodity prices plummeted, bringing farm prosperity to an abrupt end. In 1921 all four banks in Burley failed. "Times were very hard then," noted one writer. "Hay sold for $4.00 per ton and 3,000 bags of potatoes could not be sold at any price. The white fly took the sugar beets after they had been thinned and hoed."

At that time there were no local potato dealers or other agricultural services, remembered Ernest Handy, whose father had homesteaded on the Minidoka project. "People had to take their potatoes to town and put them on railroad cars. They took their chances as to whether they would get any money for them."

Many farmers lost their land. "I remember one guy—this was in about '23 or '24," Ernest said. "He got word that the bank was coming to foreclose on his place. He moved all his cattle and other livestock onto the road—I helped him pen them in. That way the bank could take his place, but he would save his livestock. The bank didn't foreclose. They could see that he had one up on them."

In the midst of this agricultural depression, Oakley reservoir filled. Although it was empty in the fall of 1920, a heavy snowfall in the winter and seven inches of rain in the spring filled the reservoir to capacity in June 1921.

"Oakley Enters Golden Age of Agriculture" proclaimed the *Oakley Herald*, while Oakley merchants advertised $pillway $pecials.

"A little water ran over the spillway last evening," reported the *Oakley Herald* on June 10; "a continuous stream began this morning about six o'clock. . . . The spillway is taking care of the water nicely, and everything is in fine shape. Farmers are jubilant."

Water ran over the spillway for a week. It was a deliberate spill. This was the canal operators' first chance to see how the spillway would work.

Water flowing over the spillway June 9, 1921. (Photo courtesy of the Oakley Canal Company.)

"It did not work that well," said Grant Sullivan, whose father later became water master. "It did a lot of damage to their canal structure. They decided they would not let the water go over the spillway again, but would let any surplus go out through the tunnel."

People in Burley were concerned about water running in Goose Creek. "Flood Waters of Goose Creek May Do Damage," announced the *Burley Bulletin*, which noted that some farmers near Burley had filled in the channel. "In fact, the Goose creek channel is practically eliminated thru this project. . . . Goose creek channel just south and east of Burley will not carry more than 10 second feet."

Despite residents' fears, Burley was not flooded. The water was dispersed into the canal system before it reached town.

Nor were full water rights enough to usher in a "golden age of agriculture." The economy remained depressed. The reservoir did not fill again.

"There are a few sticking around the flat [at Churchill] too poor to get off," reported the *Oakley Herald* April 18, 1924. "But we say, 'Perk up; there may be a ray of hope.'"

"Foreclosure seems to have attacked Golden Valley once more," it noted on August 22. "If wheat reached $2.50 cwt., we might be able to pay taxes and live on 14 inches of water."

The stock market crash of 1929, which plunged the rest of the nation into the Great Depression, had little effect on Oakley. The western farm states had been depressed since 1921. By 1929 Oakley settlers were part of a largely subsistence economy, raising enough food for their own use and raising a small cash crop to pay for water, taxes, rent, or the mortgage.

"Nobody went hungry," said John Clark, owner of Oakley's one remaining general store. "Everybody had gardens and raised a few animals. If people didn't have money, they could bring in a beef cow and trade it for provisions."

Robert Reed operated a creamery in Oakley. "Those cream checks helped a lot of people," John remembered. "It gave people a little cash."

By the late 1920s only three families owned farms at Churchill and Golden Valley. All the other families had lost their lands to the bondholders. Many had left the project; others remained as renters and tenant farmers.

"Times were hard," said Viva Poulton Warr, whose family farmed at Churchill. "But we were just kids, we didn't know how bad it was."

Sometime during the 1920s her father bought his first car. He paid cash for half of it and planned to pay for the other half when he sold his beans. But before he could harvest, the bottom dropped out of the bean market, and he could not sell his beans for any price. At the end of the year he had to give up his car.

"That was the first time I realized that we were poor," Viva said. "Dad felt so bad about losing that car. He vowed he would never buy anything on credit again, and he never did. To the end of his life he paid cash for everything he bought."

Grant Sullivan's father and uncle were among those who came from the Midwest, cleared and settled land, and then lost their farms. Grant's uncle went to Bremerton, Washington; Grant's father rented land in Golden Valley.

"We lived good," Grant said. "We didn't know we were poor, or that there was a depression. We had plenty to eat. We had a garden, a milk cow, and lots of meat—pigs, chickens, and lambs. We sold cream and got a little money from the creamery."

But he knew there was a drought. Their house was located at the bottom end of their property. Their yard had no grass or flowers because the irrigation ditch was too far away. All they had was dirt.

One year Grant's father raised a little patch of "real good potatoes" which he sold for a good price. He put the money in a bank in Twin Falls, took his five boys to town and bought them new school clothes. "A week or two later, here come a man in fine-looking clothes driving a fine-looking car," Grant said. "We couldn't imagine what anyone like that was doing at our place. It turned out that Dad's check had bounced. The bank had gone broke."

Lida Martin Hill remembered how hard her parents worked. "Dad was up and down all night to change water, to save it. Mother was up and down all night taking care of kids." Her mother made all the children's underthings out of flour sacks. Lida's bloomers had "Red Rose Flour" written on them. "But it didn't matter because everybody was the same. We didn't know what rich and poor was, especially not rich."

People had a hard time making money to pay taxes and the mortgage, remembered J.J. McLaws. One year his father had to sell a water share to pay taxes. He received fifty dollars for one share at a time when land was going for a dollar an acre.

"That was what people did," J.J. said. "They sold water and tried to make a living another way. They cut down and cut down on their farming." Sometimes they rented water from farmers who had extra at the end of the season. J.J.'s father was a ditch rider so he knew who had extra.

There were no fights over water, J.J. remembered. "The ditch rider took care of making sure everyone got their share. The headgates were padlocked, and nobody fooled with them."

When there wasn't enough water, "people would neglect their alfalfa fields," said J.J. There were thousands of black widow spiders in those dry alfalfa fields. He and his friends had fun catching black widows in a bottle. "My mother got pretty mad at me when I'd bring one of those bottles in the house," said J.J.

During the dry years people farmed only part of their land, remembered Viva Warr. "They raised one crop of hay and a little grain. They'd irrigate once and that was all. By August they'd be out of water."

They had short water years throughout the late 1920s and early 1930s. In 1931 farmers received 6 inches of water per share. In 1932 the drought forced Robert Reed to close his cheese factory because there were no longer enough dairy cows on the project. In 1934 farmers received 4½ inches of water per share.

Conditions improved after 1934. In that year the Bondholders Protective Association agreed to sell the reservoir and irrigation system to the Oakley farmers. The price was $250,000. Although this was about one-eighth of the original cost, it was considered fair. The bondholders had invested in a project that was supposed to deliver 1½ acre feet of water to 44,000 acres. They sold a project that in the past ten years had delivered less than one acre foot of water to 21,000 acres. They had bought in a time of prosperity, they sold during the depression. "It was one heck of a lot of money at that time," Winslow Whiteley said. The farmers anticipated paying on their loan until 1968.

The newly organized Oakley Canal Company announced that no more water would be delivered to Churchill and Golden Valley after 1934. All water would go to the farmers on the original Oakley project and each farmer's shares would be doubled. While the farmers would still receive 1½ acre feet of water per share with full water rights, they now owned twice as many shares. They could use two shares of water for each acre of land.

The three families who owned farms in Churchill and Golden Valley were given farms near Oakley. Most renters and tenant farmers left the area. An auction was held at Churchill and Golden Valley where everything was sold "that was worth selling that people could carry away," remembered Viva Warr. The Churchill and Golden Valley school buildings were moved to Declo.

"Dad was in charge of selling everything," said Grant Sullivan. "A lot of houses were moved to Rupert. There were no good barns. There was some fencing that was sold. People bought land for twenty-five cents an acre. They didn't live on it. A couple of sheep outfits used it to run their sheep."

The Sullivans moved to Oakley where Grant's father worked as ditch rider and water master. "I don't know how my parents felt about having to leave Golden Valley, but I thought it was great," Grant said. "It seemed so good to get away from all that dirt."

The Poultons and the Martins were among the last families to leave. "Dad was glad to get out of there," said Lida Martin Hill.

Gradually life became easier for people in the Oakley valley. As the Great Depression ended and World War II began, prices for farm products rose. High prices coincided with increased precipitation. Beginning in 1942, Oakley irrigators had five years of full water rights. In 1947, twenty-one years ahead of schedule, they paid off their reservoir bonds.

They also began to drill deep wells. In 1947 Julion Clawson brought oil-drilling equipment from Oklahoma and dug the first deep well in Minidoka County. Other farmers drilled wells on the Oakley project and on the desert lands between Oakley and Burley. During the 1950s and early 1960s Burley nearly doubled in size, its population increasing to about 9,000 inhabitants. Deep wells and sprinkler irrigation made farming possible at Churchill, Golden Valley, and Kenyon. As productive farms were established from Oakley to the Minidoka project, the dreams of the early settlers were finally realized.

By 1983 little remained of the old Goose Creek channel. As farmers expanded their operations into the desert lands, they began covering it.

Winslow Whiteley was one of the first to fill in the channel. "I owned a lot of it," he said, "and I filled a lot of it in, though I knew that it might later create a problem. Goose Creek was a weed patch and a rodent haven. I leveled it and farmed it."

Most farmers filled in the channel without thinking of future problems. Glen Jensen, county executive director for the Agricultural Stabilization and Conservation Services (ASCS) in Burley, leveled the channel on his farm without giving it any thought.

Although Viva Poulton Warr and her husband knew that the Goose Creek channel ran through the upper part of their farm, they were not concerned. "We never thought about it," Viva said. "We never even thought to think about it."

Lawrence Burch, who lived south of Burley, remembered the Goose Creek channel as a dry, weedy depression, a good place to play. He hated to see it filled in because he liked to hunt rabbits and pheasants there. "People used to say in a joking way, 'What if the dam runs over?'" he remembered, "but they were only kidding. Nobody thought it would happen."

In Burley, homes, businesses, a cemetery, and a school were built on the old Goose Creek bed. People moved into houses without knowing they were in the channel. Many residents of the Goose Creek Manor apartment complex did not know why it had that name.

Julie Fairchild, who purchased a home on East 17th Street in Burley, did not know at first that it was in the channel. "I was doing yard work and noticed the dirt was clayey," she said. "My neighbors told me, 'That's Goose Creek dirt.'"

By 1983, according to law, the Goose Creek channel no longer existed. In the early 1970s the city of Burley used it to divert water after a sudden storm. A property owner living on the channel was flooded and sued. In 1974 the Idaho Supreme Court declared that because Goose Creek had not carried water from the reservoir since 1921, and because it had been altered or eliminated in places, it no longer existed as a natural channel. The city had no right to divert water into it.

Many people were unhappy with the ruling. They knew it left Oakley reservoir without an outlet. "If Oakley reservoir flooded, the water would flow down the old channel," said attorney Bill Parsons. "The court ignored the law of physics. The old channel is the low point in geography from Oakley to Burley. It's still the natural drain."

"The judge seemed to think that his powers gave him the right to take over the powers of nature," said Oakley Canal Company director Daren Critchfield.

"It was like filling up a bathtub and not having a plug to pull out," said irrigation specialist Earl Corless.

Other people were not concerned. Although Burley Stake President Boyd Poulton had lived in Cassia County most of his life, and as a real estate broker regularly dealt with water when he sold farms and ranches, he did not know where the old channel was.

"The channel was not common knowledge to property owners," he said. "It did not decrease property value." He helped develop a subdivision on 27th and Fairmont in the old channel. "We bought flood insurance because it was near a large canal, not because of the channel."

There was some discussion about building Mountain View School, Burley's newest elementary school, in the Goose Creek channel. It was something they seriously considered, remembered Norman Hurst, who was assistant superintendent at that time. The court case declaring that Goose Creek no longer existed had occurred shortly before, "but even without the court ruling we probably would have built there." The school district owned the property. It was within the city limits so the school could use the city's water and sewer system and have other services. The fact that the school would be located on a dry channel where water had not flowed for fifty years did not seem to be important. No one thought there would be a flood.

Larry Adams on a snowbank at Bostetter, spring 1984.
(Photo courtesy of Larry Adams.)

Chapter Three

Winter, 1983–1984

No one thought of a flood in November 1983, even though 1.7 inches of precipitation for November was double the monthly average. Eleven inches of snow fell in Burley on the third of December.

The heavy snow meant long hours for Ray Kimber, supervisor of the Oakley Highway District. He was responsible for maintaining 275 miles of roads near Oakley and in the nearby mountains. "We had a terrible time keeping the roads open," Ray said. "I had the biggest crew I ever had in my life—hired a lot of extra people. It was a long, hard winter." The snow on the Goose Creek road above the reservoir was 3½ to 4 feet on the level, the most he had ever seen.

Shortly after Thanksgiving, ranchers Mike Poulton and Scott Bedke (Bed′ key) flew over the range looking for cattle stranded in the mountains. Larry Adams and his son Kelly attached a sleigh to their snowmobiles and hauled in hay bales. "I'd seen deeper snow," Larry said, "but I'd never seen snowdrifts that low before. The deer came down to the haystacks for the first time that year." By mid-winter a thousand head of deer were devouring the stacks at Kelly Poulton's ranch three miles south of Oakley.

"The snow came low initially," said Leo Busch, hydrologist with the Bureau of Reclamation regional office in Burley. "It was an early, cold winter, so the precipitation fell out at lower elevations than it usually did. At first we lacked the heavy stuff in the higher areas."

Snow covered the ground December 14 when the Oakley Canal Company board of directors held their meeting. Canal board president Winslow Whiteley, who was seventy-three years old and had lived in Oakley all his life, had never seen so much early snow.

Since 1934, when farmers had purchased the reservoir and canals from the bondholders, the Oakley Canal Company board of directors had supervised the irrigation system. They attended one meeting a month and received four dollars a month in salary. They decided how many inches of water would be allotted per share, a critical decision when water in the reservoir was low. But allocating water had not been a problem in recent years. For the first time in the reservoir's history, farmers had received full water rights for fifteen consecutive years. The directors kept surplus water in the reservoir as carryover for the next year. During the dry years of 1977 and 1981 the carryover made it possible for them to give full water rights.

The carryover had caused some concern during years of high runoff. In 1975 the carryover plus runoff totalled 112,000 acre feet which was 37,500 acre feet more than the reservoir's capacity. The directors had turned water into the canals early that year offering it free to anyone who would take it. By the time spring runoff came, they had dispersed enough water to avoid a flood.

They would need to do the same in 1984. They had begun the water year in October with a carryover of 35,000 acre feet. By mid-December the reservoir held 40,000 acre feet.

"There has been an unusual amount of snowfall in the hills for this time of year," wrote secretary Klyda Wake in the canal company minutes. "The board discussed ways of keeping close check on this. . . . Was suggested the Soil Conservation Service be contacted and the Bureau of Land Management (BLM) contacted about releasing some water across their land if necessary."

For years a little water had run down the West Canal and onto BLM land from a small leak in the dam. Although watermaster Jay Gorringe (Gor′ rinj) could send some additional water there, he could not send much. Water does not run in canals in winter the way it does in summer. It would run a little, then freeze and create an ice jam. Water behind the ice would then overflow and flood people's property.

Three members of the Oakley Canal Company board of directors at the spillway, spring 1984. Left to right: Daren Critchfield, Bud Robinson with dog Alex, and Winslow Whiteley. (Photo courtesy of Duff Woodhouse.)

Jay had been watermaster since 1977 when he had returned to Oakley after retiring from the army. "When he retired, we flipped a coin to see whether we would go to Idaho or to Ohio where I was from," his wife Loreen said. "Idaho won."

Being watermaster was a demanding job, Loreen remembered. "If Jay gave people more water than they were entitled to, they never said anything. But if the water was short, he sure heard about it."

Though Jay released all the water he could onto BLM land, "he was not that concerned at first," said Loreen. "I remember him saying if the ground is open and not frozen in the spring, the water will run into the ground."

But Kent Foster, District Conservationist with the Soil Conservation Service (SCS), was worried. The SCS monitored snowpack and predicted spring runoff. On December 13, during the monthly meeting with local farmers who comprised the SCS Board of Supervisors, Kent discussed flooding from Oakley Reservoir.

Later, Mike Cranney, an Oakley farmer and member of the SCS Board of Supervisors, remembered that although they talked about the reservoir, he was more concerned about Cottonwood Creek. Mike knew that a heavy snowpack could cause Cottonwood to flood. With no dams to hold back the water, Cottonwood flushed fast—all the water came down in one big rush. A couple of years earlier Cottonwood had flooded some farm land west and south of Burley. "Old-timers" could remember when it ran in a natural drain west of Burley to the Snake River. It would likely run there again.

The first of January the SCS issued its snow service bulletin. Precipitation was well above normal, it noted, "with amounts as much as 300 percent above average for southern Idaho. . . . "Oakley Reservoir, south of Burley is expected to fill to capacity for the first time since 1921."

December precipitation in Oakley was 1.78 inches—a new record. Total precipitation from September through December was 5.71 inches, almost double the normal of 3.17 inches.

"Oakley Dam may spill water," announced the *South Idaho Press*, the local Burley paper, on January 7. It reported that snow accumulation was about 240 percent of normal.

Most people did not understand the implication of those figures, but irrigation specialist Earl Corless knew what they meant. He knew that runoff 150 percent of normal would cause a flood.

Buddy Bedke, whose ranch on Goose Creek was fifteen miles above the reservoir, also knew there would be a flood. "I knew from the seventh of November when it set in on us and never give us a breath of air," he said. "I had snow 3½ to 4 feet deep in my front yard. Ordinarily there'd be a foot of snow there for a few days. My place is in a little valley, where the snow generally comes and goes. That year the wind would blow and the snow would come. I had to dig my hay stacks out, make roads and feed grounds, bare off a little ground for the calves. I had leased a big Case tractor with a dozer blade. I come down toward Oakley helping to keep a road open for the school bus. I tried to make sure everybody on Goose Creek could feed their cows, too.

"By the first of January I started making statements," Buddy said. "Everybody in Oakley thought I was nuts. I told them they could be in for all kinds of heck here in the valley."

Buddy's grandfather had started the ranch in 1927, and Buddy had lived there all his life. "When you live in a ranch up in this country as long as I have, you get feelings, premonitions," he said. "Somebody was telling me—maybe it was experience—I don't know what. But I knew we would be in trouble."

Government agencies began making plans. On January 5 the Idaho Bureau of Disaster Services called a special meeting to discuss spring flooding. Darrell Waller, coordinator of disaster services for Idaho, advised everyone to stockpile sandbags.

But for most people, major flooding was more a topic of conversation than a serious threat. "There was a lot of coffee-shop talk," remembered Lawrence Burch. "Everybody knew there were big drifts. They were trying to estimate the water content. Somebody else said, 'That's all baloney. Oakley reservoir will never fill.'"

"They kept measuring the snow," said Oakley resident Suzanne Critchfield. "We'd get reports. We could see the snow. Hilton and I were so excited about the moisture. We had no concern that there might be too much."

"We thought it was a great year," said Viva Warr. "We knew there'd be plenty of water."

Everyone relied on the SCS for snow information, reflected Kent Foster. This year the SCS had asked Reeder Flying Service to collect data from non-automated snow courses. Kent thought they were giving him accurate information, but in this critical year he had to know for sure. He assigned two SCS employees to go by snowmobile to check.

The employees found that the Reeder Flying Service reports were accurate. But they also found something more troubling—extremely high moisture content in the snow on the lower courses. The lower stuff was important, because it would all run off at once. But the SCS gauges were located at higher elevations, 6,500 feet or more, and all SCS forecasts were based on the high-level snow. "Data is only as good as the way it is collected," said Kent. In this critical year, they had no data on the low-level snow.

It was cold in January with high pressure bringing fog and temperature inversions. The cold, dry weather continued the first two weeks of February. When Kent Foster met with the canal company directors February 15, he told them that while December 1983 had been the wettest December on record, January 1984 had been the driest. If they received average precipitation for the rest of the year, the SCS now predicted runoff 164 percent of normal.

Stephen Bywater, the new canal company lawyer, attended the February meeting. Because Steve had been in the area only since 1976, he knew nothing about the reservoir, the dry years, or the Goose Creek channel. He soon found out.

"The first thing the directors mentioned was the high snow pack," he said. "They were afraid there would be too much water. I was asked to research what their liability would be if the reservoir should spill. 'This will never happen,' they told me, 'but just in case it does, we'd like to know.'"

Steve made his report at the meeting. The safest legal route, he told them, was to do nothing. It was their right to let water go over the spillway and down the creek. But though the directors understood their rights, "there was a strong feeling that it was irresponsible even if it was legally correct," remembered Steve.

A big storm hit the day after the canal company meeting, closing schools and bringing the monthly precipitation to .65 inches, .02 inches above normal. Then cold weather and temperature inversions settled in again.

Since the flurry of newspaper articles in January warning about Oakley Reservoir, the *South Idaho Press* had carried no more reports about a possible flood. "Water outlook remains good," it announced the first of February. "Snow situation unchanged," it reported in March.

Assistant civil defense director Terry Bingham was not thinking about flood the first week of March when he and Bud Hoffbuhr traveled by snowmobile from Oakley to Magic Mountain. Conditions for snowmobiling had never been so good. Both he and Bud were amazed by all the snow—much deeper than either of them had ever seen it. Although they traveled along the drainage area of Oakley reservoir, Terry did not think about the significance of the heavy snowpack. Later he remembered noticing how much snow was in Bostetter. At the time, however, he did not know that Cottonwood Creek drained out of Bostetter. For him the huge snowpack meant a day of fun.

The snow did not mean fun for Ennis Pickett, canal company director and part-owner of Pickett Ranch and Sheep Company. Born in Oakley in 1919, Ennis had been farming and ranching as long as he could remember. In all those years he had never seen snow so deep, especially in the low foothills. When he put covers on his lambing shed the end of January, he found a bucket with seven inches of ice in it. That indicated to him how saturated the ground was underneath the snow. He knew water from melting snow would run off instead of percolating into the ground. He also knew that a half inch of runoff over the entire Goose Creek watershed would be enough to fill the reservoir.

On his Goose Creek ranch, Buddy Bedke was trying to protect his six hundred newborn calves. The snow on his ranch had turned to ice. "When the calves would lay down, their body heat would melt the ice around them," he said. "They'd be laying in the ice in little pockets three to four inches deep. We'd be on horses twelve to fourteen hours a day trying to get those little calves up to keep them from freezing. We'd try to move them to a hill, under the cedars, wherever we could find a little dry spot. I called Tom Blayney, a veterinarian in Burley, and had him come up to tell me what to do. He said, 'There's nothing to do except what you're doing.' We lost about fifty calves. It was not a problem financially, but it made me feel so sad. It's something I'll always remember—those little calves laying there slowly freezing to death."

The cold, dry weather continued into March. During some years farmers would have been working in their fields; this year the ground was covered with snow. It did not begin to thaw until the middle of the month.

"March 19. Typical March weather, windy, sleeting and cold," wrote County Commissioner John Adams in his journal.

"Trying to get tractor started," he noted March 21. "Concerned about fall grain."

"March 24. Rained, skiff of snow. Had it about every day.

"March 27. Woke to snow and cold. Cut down a tree because we couldn't farm."

The canal company directors held their monthly meeting March 15. "The board again discussed the water situation and plans that could be implemented in order to keep the reservoir from spilling," wrote secretary Klyda Wake. "Mr. Bywater has been doing some researching on it and Mr. Gorringe reported on plans he had in mind."

Board members approved Jay Gorringe's request to install extra headgates in order to run more water onto desert land. They talked about putting water in gravel pits. Winslow Whiteley announced that he had arranged for the directors to meet with the State Department of Water Resources and the Bureau of Reclamation March 21.

The Bureau of Reclamation supervised Minidoka Dam, the Burley Irrigation District (BID), and other irrigation systems on the Snake River. Although Oakley reservoir was not part of the Bureau of Reclamation system, the officials were concerned about it. They knew flood waters from the reservoir could ruin the BID canals.

"In 1964 or '65 a lot of water from the North Side broke the Minidoka irrigation canals," said irrigation specialist Earl Corless. "In the little time I had, I put six-foot culverts next to the four-footers already in place. The water hit and washed them both out. It ate up canal banks, took out structures half as large as my office. Water got into people's houses—people were driving in ponds. There were lakes of twenty to thirty acres. Water can go anywhere. When it gets to running, it's powerful and dangerous."

The flood had caused several hundred-thousand dollars' damage to the irrigation system. "But that would be one-tenth of one percent of the damage of a flood from Oakley reservoir," Earl said.

Leo Busch found some old flow charts which showed runoff for Goose and Trapper creeks since 1919. He and Earl examined them. The charts showed a history of sudden, rapid runoff. Leo checked information stored on the computer, trying to find a similar year. If he could find something similar, he could better predict what might happen. He could not find anything. Because they lacked data for the low-level snowpack, there was no historical basis for determining how much water might come down.

At the meeting with the canal company officials Leo was impressed with the directors' handling of the situation. "They were trying to flag the problem, trying to get everybody to look at it," he said. The directors talked about the releases they were making, and the difficulties of increasing releases because of damage to fields. They discussed putting water in gravel pits. Someone mentioned running water to Murtaugh Lake.

Leo advised them to get rid of as much water as they could in any way they could. Time was critical. By releasing a small amount of water over a long period of time, they might prevent having to release a large amount later.

"If only I'd had a crystal ball, I'd have let more water out in September," lamented watermaster Jay Gorringe.

At the SCS office, civil engineering technician Burke Scholer also worried about flooding. Burke had lived in Burley since he was seventeen and had worked for the SCS for nearly thirty years. From 1957 through 1977 he had gathered data from the snow courses. He knew from his twenty-years' experience that there had never been a winter like this one. He was certain the reservoir would spill.

Burke looked at topography maps, trying to find where the water would go. The water would seek the lowest point which was the old Goose Creek channel. County roads now bisected the channel almost every mile; between the roads were concrete ditches and other obstructions. Water would run in the channel until it hit an obstacle, then it would back up, spread out, break through the obstacle, rush to the next obstacle, spread out, and break through again. It would break the BID canals. Designed for irrigation, the canals could not control flood waters.

The water would hit Burley. It would flow in the vicinity of the old channel, Burke projected, flooding homes, schools, the cemetery, the shopping center. Burke wondered where the water would enter the Snake River. The channel split after passing the shopping center; over the years Goose Creek had taken several paths to reach the river.

Burke looked for the lowest place where the water could cross the highway and railroad tracks. He found that the lowest place was not in Burley at all, but a couple of miles to the west near the

Ore-Ida potato processing plant. Burke remembered that area. It was a natural drainage where Cottonwood Creek had once flowed. Would it be possible to divert the water into the old Cottonwood drainage?

An abandoned Union Pacific railroad track ran north of Oakley close to the Goose Creek channel. "It was easy to go down the railroad," said Burke. "Then once we were in the natural drain it was free sailing." The difficulty would be in getting from the railroad track to the natural drain. A survey would be needed since the USGS maps did not show the small details. Still, it was a possibility. Although a diversion canal would damage fields, it would be less destructive than letting the water flow through Burley.

Burke began talking with friends, neighbors, and anyone he saw. Something needed to be done, he told them. Flood waters from Oakley reservoir could cause massive destruction if the water ran uncontrolled.

Burke alerted Kenneth Fronk, a member of the Burley City Council, whose house in south Burley was a hundred yards west of the Goose Creek channel. Ken was not concerned. The reservoir was more than twenty miles from Burley. He thought that if water did go over the spillway, it would spread out, percolate into the ground, and dissipate long before it reached the city.

Other people besides Ken had difficulty imagining a flood. J.J. McLaws worked for the SCS, saw the snowpack data, heard Kent and Burke discuss the problem, and still did not think the reservoir would fill. J.J. remembered growing up in Oakley during the dry, depression years. He had played on the reservoir tower when he was a kid. He bet a friend fifty dollars that the water would not run over the spillway. "I figured it was a safe bet," he said. "The odds were on my side. It hadn't run over for sixty years."

But Mike Cranney listened to Burke. He talked to the canal directors about making an emergency channel. It could be done, he told them. The SCS officials had located a route. "Have you talked to the landowners?" one of the canal directors asked. "You'll never be able to get easements."

"Building a diversion channel in March would have been the smart thing to do," Mike later said, "but it wasn't practical. It did not take human nature into account."

Most people did not believe there was a problem. The farmers would not have allowed a canal to cut through their land, and no one would have been willing to pay for it. They needed government permission to do it and an environmental impact statement. The danger would have come and gone while they were filling out papers.

Then too, there was the possibility that with good weather, major flooding could be avoided. In 1975 the directors had averted flooding; the conditions in March 1984 did not appear much different from those of 1975. There were 50,000 acre feet in the reservoir in March 1984; there had been 46,000 acre feet in March 1975. The SCS estimated that the 1984 April-through-July runoff would be 48,500 acre feet; the 1975 April-through-July runoff had been 51,000 acre feet.

In spite of the early cold weather, March had been a good month. March precipitation at Oakley was .36 inches—.51 inches below normal. The SCS bulletin for the first of April announced that snowpack in the Oakley Reservoir drainage was 132 percent of average. If the temperatures warmed slowly and April was dry, the snow would melt gradually. With farmers using the water and Jay Gorringe sending it every place he could think of, they might be able to avoid a flood. Everything depended on the April weather.

Duff Woodhouse, Winslow Whiteley, Jim Mai, Ennis Pickett, Leon Adams, and Grant Sullivan in the canal company office.
(*Photo* The Times News.)

Chapter Four

April 1984

It snowed the first two days of April. "Really nasty," wrote Dorothy Henderson in her journal. "The daffodils on the south side of the house are all in bloom in the snow." On April 5 snow survey coordinator Jerry Beard told the *South Idaho Press* it was "highly likely" that Oakley reservoir would fill.

Jay and Loreen Gorringe drove to the Goose Creek measuring gauge. "Jay was checking it oftener and oftener," said Loreen. "Spring runoff hadn't even started. That was the first indication I had of lots of problems."

On April 6, 10, and 12 it rained—.58 inches total precipitation. Daytime high temperatures were in the forties and fifties; nights were below freezing.

During other years of surplus water farmers irrigated in early April. This year it was too cold and wet. Many years they used sprinklers in mid-April. This year it froze every night. Though sprinklers could run during a light freeze, building icicle patterns on fields and fences, sprinkler pipes would break if it were too cold. No one was willing to risk damaging a half-million-dollar irrigation system by running water during freezing weather.

On April 10 Jay Gorringe received a letter from David Hollingshead, director of the dam safety section of the Idaho Department of Water Resources, warning about possible flooding. Allen Merritt, engineer with the regional dam safety division, called Jay. "I had inspected Oakley dam and knew that the channel below was being farmed," Allen said. "The dam itself was not in any jeopardy, but it would tear the heck out of everything if the water went uncontrolled over the spillway."

Oakley Highway District director Bill Cranney, still more concerned about flooding from Cottonwood Creek than from the reservoir, selected a minor road in the area where he expected Cottonwood to come and showed it to manager Ray Kimber. "I took Kimber and said for him to be ready when I gave the signal," Bill later said. "We'd need to take out the road. I wanted him to take out all culverts and spot mainlines. I was planning on building a good-sized ditch to take the water into the brush. I knew it would not be enough, but at least it was a start."

On his dairy farm south of Burley, Oakley Stake President Rex Gerratt was also making plans. All winter he listened as people talked about the heavy snowpack. He had seen the snow himself when he rode the small commuter plane from Twin Falls to attend dairy board meetings. Since January stake emergency coordinator Wayne Mullen had been reporting on flood possibilities during High Council meetings. Though most people could not imagine Oakley ever having too much water, Rex knew there would be flooding in his stake area. The people must be prepared.

The stake and each ward had an emergency director, but because emergencies did not happen often, directors usually did not do much. This year emergency directors would play an important role. At a meeting with ward bishops, Rex told them, "We are going to have a flood. If you want your emergency director to protect your house, leave him in that position; but if you do not think he can do it, release him and get someone who can."

In Burley, Undersheriff Billy Crystal had reached a decision. Because he intended to run for sheriff, and because the primary election would be held in late May when flood danger would be at its height, he decided to resign as civil defense director. He did not want to campaign and fight floods at the same time. He knew that assistant director Terry Bingham was qualified to handle emergencies. If Terry were civil defense director and Bill the assistant, they could continue to work together; but Terry, not Bill, would be in the limelight. Terry and the commissioners agreed to the change, making it official at the commissioners meeting April 9.

Oakley Canal Company watermaster Jay Gorringe on the old gates that divert the reservoir water into the Goose Creek channel, spring 1984. (Photo courtesy of Ryan Thomas.)

In his new position Terry decided to investigate the rumors he had heard about Oakley reservoir. He found Jay Gorringe and Rex Martin at the Basin cemetery, using the canal company backhoe to dig a grave. He chatted with them a little.

"Lots of water up there in the hills."

"Yes, no one had ever seen anything like it before."

"Could the reservoir hold it all?"

"Probably not."

There was carryover left in the reservoir from last year, plus what had been added from winter and spring storms, Jay told him. They were trying to get rid of it. They had been sending water to BLM ground all winter. They were encouraging farmers to take early irrigation water and turning streams of water into gravel pits around town. It wasn't enough. The moisture in the snowpack would still be greater than the remaining capacity of the reservoir.

"Where is the water going to go when it gets to the top?" Terry asked.

"To the Snake River," Jay replied.

Terry and Bill met with the canal company directors April 18. Though the directors had talked earlier about flooding, they had always hoped that with the right weather they could handle the water. The cold, wet weather dimmed that hope. It had warmed a little—on April 14 to sixty-eight degrees, to seventy-seven degrees April 16. Though nights were still cold with temperatures near freezing and though daytime temperatures had again dropped to the fifties, Jay Gorringe filled the irrigation canals.

"Mr. Gorringe, watermaster, reported on the water presently being released and needs to start taking out more," wrote secretary Klyda Wake in the minutes. "Mr. Terry Bingham was asked to contact Kloepfer (Klep′ fer) Concrete Co. for permission to spill water into their gravel pits north of Oakley if necessary. . . . A motion was made by Daren Critchfield that a meeting of stockholders be called."

"Oakley Dam near flood stage" announced the *South Idaho Press* headline April 18, but the article itself was more reassuring than the headline. "Oakley canal company officials are looking for places to put surplus Goose Creek water as they try to build a safety margin between the level of their reservoir and the spillway at Oakley Dam. . . . Oakley Canal officials say the storage space should be more than enough to handle the remaining spring runoff under normal conditions. . . . 'We've got our fingers crossed,' said Winslow Whiteley, chairman of the Oakley Canal board of directors. 'We're spilling water everywhere we can right now, and I think we're going to be all right.'"

Winslow did not want to alarm people. Besides, he had lived through so many dry years, that in spite of the heavy snowpack, it was hard for him to imagine there could be a flood.

That afternoon the directors received a letter from former Burley resident Dal Wayment, a canal company consultant who worked with James M. Montgomery Consulting Engineers in Salt Lake City.

"I have prepared a water budget for the Lower Goose Creek Reservoir based on runoff information supplied by Mr. Jerry Beard of the Soil Conservation Service in Boise," he wrote. "The lower snow pack is at about 200% of normal. Because of this Mr. Beard feels that there will be a heavy early runoff. He is estimating that about 31,500 acre feet will runoff during April and May. An additional 17,000 acre feet should runoff during June and July. These estimates are based on 'normal' precipitation for the rest of the season."

Dal gave the estimated inflow for March through September along with the water releases for 1983, concluding that there would be 8,000 acre feet of water above the capacity of the dam. He suggested releasing 12,000 to 16,000 acre feet as a safety factor, adding, "It will require a flow of 135 cfs for 30 days to release 8,000 acre feet. This would be in addition to amounts normally released for irrigation."

Eight thousand acre feet would cover 8,000 acres with water one foot deep. There were 10,500 acres in the entire Oakley project.

Cool, wet weather continued the last two weeks of April. A blizzard hit the night of the twenty-fourth. Wind and snow whipped through the area as the temperature plummeted. The next day two inches of new snow fell; the high temperature was 34 degrees.

"April 25. Ground blizzard all day," wrote County Commissioner John Adams. "The snow didn't really light until afternoon. The snow blew all the canals shut and turned the water to ice."

The next morning John found the ditch by his field full of ice and snow. The canals were frozen shut with water running everywhere. John found Hilton Critchfield stuck in a snowdrift and pulled him out.

"Two days of snow," wrote Dorothy Henderson April 26. "Trees—ground all covered white. The tulips and grape hyacinths are all in bloom amid the snow."

Precipitation in Oakley was .31 inches for the two days; two more feet of snow had fallen in the mountains.

On the night of the blizzard the Oakley canal company shareholders' meeting was held. Winslow Whiteley and Jay Gorringe described the reservoir problem and asked for suggestions to keep the water from spilling.

Several people made suggestions: Put water on fall grain, scrape BLM ground to find gravel spots which would soak up the water, send it down the old Goose Creek channel, build a ditch along the railroad tracks, give everyone some water to take care of. Someone suggesting dumping water into old wells—that might be a problem with the environmental protection agency. One person suggested raising the spillway—they would have to get permission from the Department of Water Resources—it could jeopardize the integrity of the dam.

But though there were many suggestions, no one could think of a good way to dispose of a massive amount of water. Many did not think such a course was necessary. "I had a lot of skepticism about what they were saying," said Oakley farmer Randy Hardy. "We knew there was a lot of snow, but we want lots of snow. They said we need to do all we can to run water, but we had trouble running sprinklers because of the ice factor at night. My biggest feeling was that there weren't any good answers to it."

"There were really divided feelings among the people," remembered Mike Cranney. "Some were saying, 'We've had drought all these years. We finally get a full reservoir, and now you want to let it go.'"

"All these Oakley farmers get together and there is little concern about the dam, just concern about past grievances," wrote John Adams in his journal.

But Ennis Pickett was scared, John remembered. "The dam's going to run over," Ennis told the group. "There's so much snow up where we run our sheep. I've never seen anything like it."

"Wanted: Good home for Goose Cr. water," proclaimed the headline in the *South Idaho Press*. "If something isn't done soon, it's going to go over, and it's going to cause a lot of flooding. That's my opinion anyway," said Bud Robinson.

Yet the feeling of unreality persisted even among those who understood the problem. "I never thought we'd ever see the day when we'd be looking for ways to waste water," said Daren Critchfield.

While canal company officials hunted for places for water, Oakley Stake President Rex Gerratt told LDS church leaders in his stake to prepare for a flood. He and stake emergency director Wayne Mullen had marked the reservoir banks so they could tell at a glance how quickly the water was rising. They updated their lists of persons with equipment, training, and resources for helping with emergencies. They assigned wards to fill sandbags. "Some people had some doubts about how serious it was," Rex said. "I told them, 'We will have a flood,' and gave them a number of sandbags to fill."

He and Wayne surveyed Birch Creek and found branches and other debris that could impede the water's flow. Rex asked two ward emergency directors to clear the branches. When they had not done it by the next meeting, he stopped the meeting, sent them to do it, and told them to return to the meeting when they were finished.

"It was critical that those things were done," he said. "I knew there was going to be a flood. There had to be an attitude among the people that we had to prepare."

County officials were also preparing. On Thursday, April 26, the day after the irrigators meeting, local civil defense leaders from southcentral Idaho met with officials from the State Bureau of Disaster Services. Usually only civil defense directors attended those meetings, but because of the flood threat, County Commissioner Weldon Beck attended also.

Ordinarily it would have been a routine meeting with state officials talking about earthquake, flood, and nuclear attack, while local directors made plans for disasters they never expected to happen. This year's meeting was different. Floods could occur anywhere in southern Idaho. Officials came with briefcases bulging with information. State officials Darrell Waller and Paul Massie had asked Bob Thuring of the Emergency Management branch of the United States Army Corps of Engineers to address the meeting. Cities and counties could request help from the Corps of Engineers which was the main federal flood control agency.

Terry Bingham made notes as the officials talked: The Corps of Engineers could provide sandbags, but these were not free. The Corps could help minimize flood damage through its advance measures program and help to determine exactly where the water would go. State funds might be available through the governor's office; there could be disaster-related individual or family grants; everyone should buy and maintain flood insurance.

Terry listed things to be considered in an evacuation: determine who would be in charge, set up safe areas, check on transportation and management of shelters. Officials could not forcibly evacuate a person, but could only give warning.

"Have commissioners request survey of dam and watershed of Oakley dam danger," Terry wrote.

Terry and Weldon sat by Bob Thuring during lunch. Bob was surprised to learn that Oakley reservoir had no outlet. "You've got a problem here," he said as he looked at the information they had brought. "You need to put together a formal request for assistance from the Corps. I'll give Tom Tate a call and have him come down and look things over."

The next day county attorney Al Barrus drafted a request for assistance from the Corps of Engineers. Though initially unconcerned about possible flooding, Al was worried now. Recently he had driven to Grouse Creek, a small ranching community located above the reservoir along a tributary of Goose Creek. "That was the first time I got spooked," he said, "to see how much water was already starting to come, and spring runoff had not even started." Though he had been to Grouse Creek many times before, he had never seen water like that. As county attorney, it was his business to keep the commissioners out of trouble. He did not see how they could possibly avoid trouble now.

On his ranch at Goose Creek, Buddy Bedke knew a flood would come soon. "The snow turned to ice, the ice to slush, then the slush to water," he said. "I'd drive with a pickup to where I was feeding calves—water and ice would spray up all around. The snow was setting under the water—ten inches or a foot deep—setting there just waiting to get warm enough to run."

The canal company directors met again April 27. "We've got to get moving right now," Winslow Whiteley told the Twin Falls *Times News* reporter who was present. "We've talked about this thing long enough."

"We're going to have 50,000 inches in our lap if we don't do something," said Jay Gorringe.

"The proposal of putting water in the Cranney gravel pits, Magic Valley Sand and Gravel Co., and the different places where more water can be dispersed were discussed," recorded Klyda

Wake. "Permission was given to spill water in the Trafford Woodhouse gravel pit, OK'd by Duff. Winslow Whiteley OK'd dumping water in gravel pit east of the Mitton place. Motion was made by Ennis Pickett that Garth Greenwell be contracted and hired to repair the Golden Valley ditch. . . . Discussions are going on with Wade Baker family and Schroeders to run west canal water to them."

Secretary Klyda Wake listened to everyone's concerns as she dutifully recorded the minutes. She was not worried. The directors had always managed to take care of surplus water in the past. She did not see why this year should be any different.

But at the SCS office, Kent Foster knew better. Kent had seen the figures for the snow courses the SCS measured. Ordinarily in April the snow water content decreased about four inches. This April it had increased eleven inches. Kent knew that no matter how hard the canal officials tried, they could not disperse enough water. The weather in April had been too cold and wet. There would be a flood. The only question was how soon it would come.

Lower Goose Creek Reservoir, April 28, 1984.
(Photo The Times News.*)*

Chapter Five

May 1–May 6

Until the end of April, only a few people had worried about the reservoir. "The emergency snuck up on everybody," said Golden Valley farmer Leland Baker. But by the end of April people in Oakley were concerned.

County Commissioner John Adams ran water onto his alfalfa field. Although earlier he had complained in his journal about public apathy, now it seemed to him that everybody suddenly realized "they needed to get in high gear."

Jay Gorringe filled the irrigation ditches, giving water to people whether they wanted it or not. "We irrigated crops to their detriment, especially the grain crop," said Kim Cranney. "They would call and say, 'You have 300 inches of water.'

"We'd say, 'We don't need any water.'

"They'd say, 'It's coming anyway.'"

Viva Warr was more worried about the irrigation water than she was about a flood. "They kept putting it in, and putting it in, and putting it in," she said. "We didn't know what to do with it."

Until the end of April, Nona Jenks had not thought much about the reservoir. She and her husband Orvin lived north of Oakley at Island, a place where Goose Creek had originally divided into two channels.

"As of last Sunday I have become concerned about a flood occurring in the little Oakley valley and perhaps Burley," Nona wrote in her journal May 2. "Unless the Almighty calms or tempers the weather, or a miracle happens, we are likely to see the Oakley reservoir going over her bank. . . . It seems spooky to think of anything like this happening. . . . In Church Sunday Bishop Cranney said we are all to do what we can or should to prevent disaster from happening to our little valley.

"So if the water goes over, where will it go? Our home is ten miles straight north from the reservoir and in a direct line, but we are on an island, up a little. We've had offers to borrow a couple of boats—Tim's and Bill's."

Loreen Gorringe realized for the first time how worried Jay was when they drove to Twin Falls one afternoon. "We'd gone the back way and we were talking," Loreen said. "Jay was saying if the dam should overflow, where would we put the water. He didn't know what to do with it.

"A long time ago there was a ditch that went out to Cottonwood. Jay said, 'Why not send it that way? Why not send it all the way to Murtaugh Lake if we have to?' We drove around that way to look at it. Jay could see that it was the perfect place."

Sunday, April 29, stake emergency coordinator Wayne Mullen, the stake presidency, and the three Oakley bishops held an emergency meeting. "We impressed on the bishops the need to get on the ball now to prepare," wrote Wayne Mullen in his journal. "They took hold and planned the following [Wednesday] night to fill sandbags."

"Each ward became a little sandbag factory," Stake President Rex Gerratt said. "People became very proficient in filling sandbags."

"They filled sandbags in our spud cellar," said View farmer Kent Searle. They dumped sand into the back of a truck that had a "tare" or dirt elevator. It carried the dirt so they could catch it in sandbags at the end. "A lot of dirt spilled, but that was no problem. We could shovel that up. It was a lot faster than filling bags by hand." They set the sandbags on pallets so they could be moved with machinery.

"The youth filled sandbags at the church," said Pella farmer David Beck. "They filled them with our potato conveyor—we fixed a spout on it. The kids would hold up a bag, we'd fill, they'd move, we'd fill up another one. Everybody really got involved. It was part of our emergency preparedness. We were given so many bags to fill."

Oakley resident Richard Hardy was amazed to see people filling sandbags. So much water was running into gravel pits and alfalfa fields; he thought that would take care of any surplus.

City officials in Burley were not worried about a flood. Mayor Chuck Shadduck and the city council talked about the threat from the reservoir, but did not think Burley would be affected.

Bureau of Reclamation officials in Burley were concerned, however. They knew that if the water ran down Goose Creek, it would ruin the irrigation canals. They talked about the court decree, which said that Goose Creek was no longer a reservoir outlet. "We don't need to worry about the reservoir," one official said. "All we need to do is take the court decree and show it to the water. Once the water sees that the Goose Creek channel no longer exists, it will know it's not supposed to run over."

Computer technician Denny Davis, who was also a counselor in the Burley Second Ward bishopric, helped plan for the evacuation of everyone within the ward boundaries who lived close to the channel. "We each took a third of the ward," Denny said. "There were eighty-four widows. I remember thinking, 'I wonder if they would mind riding in the back of my pickup?'" He hoped they would not have to evacuate, "but I had suitcases packed, ready to grab and go."

Administrators at Cassia Memorial Hospital in Burley organized a flood committee to plan for evacuation. The hospital was on low ground close to the channel. "Our responsibility was to work out details," said committee member David Gibbons. "We asked each department to make a list of things they needed to function—everything from bandages to pens and pencils—and to prioritize that list. We planned to move to the Star church where we would function as an emergency center."

Although the hospital board would make the decision whether to sandbag or evacuate, Dave felt evacuation was a strong possibility. Even if they were able to protect the hospital by sandbagging, the flood waters would isolate it. The hospital would be useless during an emergency if it were inaccessible.

The water would probably not be high enough to drown anyone, but Dave knew there would be casualties—automobile accidents as people rushed to high ground, injuries as they tried to protect their possessions. If the flood came through Burley, there would be panic, and panic was always dangerous.

"Most people, when they think of flood, think of water," Dave said. "They think of clear irrigation water getting in their basements. But floods are not just water. They're mud, debris, sticks, grass, chunks of wood—anything the water brings down." Sheet rock, wall board, insulation would be ruined in basements—furnaces, deep freezes, everything with a motor. People could sandbag, but it might not help. Most people in Burley did not have valves to keep water from coming up from the sewer.

Dave knew that his house, a few blocks from the Goose Creek channel, would be flooded. He would not try to sandbag, he decided. If he heard the water was coming, he'd load his family and valuables in their travel trailer and tow it to higher ground. He figured he would have at least eight hours from the time the water left the reservoir until it arrived at his place. But he was also Burley Third Ward Elders Quorum president, responsible for seeing that others were helped, especially the elderly and the disabled. As he thought about his family, the hospital, and his church responsibilities, he wondered what he would do first. One thing he knew—those eight hours would pass quickly.

In Oakley, canal company director Ennis Pickett had been watching Birch Creek where the runoff would come first. The creek was running high Sunday night—he estimated it to be about 10,000 miners inches. "We had visible evidence of a real problem," he said. As soon as he arrived home, he called Terry Bingham. "If you have any doubt that we will have problems with the reservoir," he told him, "come and look at Birch Creek."

The Goose Creek channel ran through Picketts' property about eighty yards from their house. It was a shallow depression which Ennis had filled in and farmed. Monday, April 30, Ennis called Warberg's in Twin Falls and asked them to store his furniture. He hauled his farm equipment east of the railroad tracks. Then he and his wife Dorothy moved to their sheep camp north of Oakley. Ennis believed that if they did everything they could to protect themselves, and then had damage, it would not matter. "But if we sat and did nothing, it would be on our conscience."

"Interest up on flood insurance," reported the *South Idaho Press* on Monday. Insurance agent Dave Kerbs warned residents to check their coverage—most home-protection policies did not cover flood. Only two flood policies were available—Lloyd's of London, which had a two-weeks waiting period, and the National Flood Insurance Program which did not cover basements.

Monday afternoon Civil Defense Director Terry Bingham met with the county commissioners, SCS officials, and canal company directors. Water will run over the spillway, Jay Gorringe told the commissioners. The reservoir level was fourteen feet from the spillway, rising an average of three inches a day. At that rate it would spill in forty-five days, but it might spill as soon as fifteen days if the weather warmed.

"We've been putting water in every hole we can find," said Jay. "As far as getting rid of enough volume of water to do any good—it just isn't there."

Commissioner Weldon Beck thought that the canal company should take care of the water. "But canal officials seemed to think that as long as the water was in the reservoir, it was theirs, but once it ran over, it was the county's water," he said.

It was not their fault, countered the directors. Goose Creek had been their means of diverting surplus water. When the channel was filled in, their only means of diversion was taken away.

"They want you to control nature and then take away your means to do it," said director Daren Critchfield. People were upset with the canal company because they had water in the reservoir. "But water in the reservoir is like money in the bank. Just because the bank is full of money you don't take it out and start throwing it away."

The Oakley Canal Company was not a government agency. The directors could not dig a ditch or dump water on people. Their only course was to let the water run over the spillway. They could be sued if they tried to do anything else.

The county could also be sued, attorney Al Barrus warned. If the commissioners diverted water out of its natural channel and damaged private property, the county would be liable. If they did nothing and damage occurred, it would be considered an "Act of God." Legally, their safest course was to let it run over the spillway.

Yet both the commissioners and the directors knew that if that happened, it would be the biggest disaster in the county's history. It would damage thousands of acres of crops and farmland and destroy the county's tax base. It would wash out the BID canals and leave farms west of Goose Creek without water. A third or more of Burley would be flooded, including Mountain View School, Overland Shopping Center, Gem Memorial Cemetery, the hospital, and probably Southwest and Dworshak schools.

Weldon feared that water running uncontrolled would wash out the south bank of the dam. Then they would have another Teton disaster, with water carrying away homes and buildings. People remembered Teton, the smashed cars and houses, the smell, the mud. In spite of liability they had to do something. They could not let a Teton disaster happen in Cassia County.

They had to do something, but what? Terry took notes as the group brainstormed: Build a channel to Murtaugh Lake. They would have to contact the Twin Falls Canal Company—there could be legal problems. Let it percolate into the ground at every opportunity. There might be crop damage and damage to irrigation lines. Run it along the abandoned railroad right-of-way to the Golden Valley road. Run it past the Marion road neighborhood. The Army Corps of Engineers could help or advise them. They would need a declaration of emergency. The telephone company would have to move underground cables. They could do some grading in the borrow pits. Paul Massie in Boise, head of the state Bureau of Disaster Services, said that the Corps of Engineers was sending Tom Tate, "the best flood fighter in the West."

The *South Idaho Press* reported the meeting. In a huge headline it announced, "Experts agree: Oakley Dam will spill."

Tom Tate was assisting with flood control on the ice-jammed Salmon River in central Idaho. Anyone involved with flood control in Idaho knew Tom. He had grown up in Mountain Home, Idaho, joining the Corps of Engineers as a surveyor in 1948. In 1963 he transferred to the Corps' newly-formed emergency management division. He had been "Mr. Flood Control" ever since. During the twenty-one years he had worked with emergency management, he had missed one Christmas, many Thanksgivings, wedding anniversaries, and high school graduations. He always had two suitcases packed, ready to leave at a moment's notice.

He had nearly completed his work along the Salmon River and was thinking of returning to Walla Walla when Corps headquarters called. "I was told to go down to Burley and have a look," Tom said. "They said they might have some trouble down there. I got the phone call Monday night. Tuesday morning I was on my way."

While waiting for Tom, Terry Bingham worked on the numerous details that needed attention. They must have more sandbags. Those ordered in February would not be enough. Someone had suggested using potato bags, since they were cheaper and readily available. Terry ordered ten thousand potato bags from Twin Falls. They could use them, he decided, but regular sandbags were better. He called Friedman Bag Company and Chase Bag in Boise. They had sandbags on hand and would send them immediately.

Terry installed two new radios in his sheriff's pickup. The pickup already had an eight-frequency radio; the two new radios had sixteen frequencies each.

Terry talked with volunteer reserve deputies Kent Searle and Leland Baker, who often helped the sheriff's department. Leland's pickup was equipped with a CB radio, a sheriff's band radio, and a portable telephone. It had portable gas tanks. A slogan along the side advertised his family's potatoes—"Baker's Bakers." Leland installed more radio frequencies after Terry talked with him. He was prepared to do whatever was needed.

Tuesday, May 1, at 3:15 P.M., Tom Tate drove into town. "He was very businesslike, very matter-of-fact, very knowledgeable," Terry said. "He spoke plainly. He had a weathered complexion that showed he'd been in the sun. He walked with a distinct limp—I imagined it was gout, as he seemed to be in pain—but it didn't slow him down any." Tom would be able to help them, Terry thought. He knew what he was doing.

Tom Tate, Corps disaster coordinator Donna Street, and state emergency coordinator Paul Massie toured the area with Terry. Although the officials could see the danger, they could also see ways of averting it. Tom did not think it would be a big problem. They would build a small canal to carry the surplus water into the

Terry Bingham talking on the radio in his sheriff's patrol vehicle. (Photo courtesy of Terry Bingham.)

desert, and let the water soak into the ground. "I figured we could get it done easy," Tom later said.

That evening Terry Bingham met with Oakley Stake President Rex Gerratt and the stake and ward emergency directors. "We had a liaison with the LDS church, with Rex Gerratt, from the beginning," said Terry. "Someone suggested to me that he get involved."

The stake Young Men's and Young Women's organizations were planning a sandbag class for the youth at Picketts' gravel pits. Terry explained procedures for getting sandbags, and passed out Civil Defense booklets describing the proper way to fill them. Sandbags should not be full or they would be hard to handle and the sand would run out when they were stacked. Leave a flap at the top to fold over, he told them.

Some people were building dikes around their houses. That could be a problem, Terry warned. If water broke through a dike, it would blow dirt from the dike into the basement, and they would have an even bigger mess. If they were going to use dikes, they needed to reinforce them with sandbags or stone.

The weather continued cool. It rained Monday, Tuesday, and Wednesday, April 30 through May 2, giving Oakley .28 inches more precipitation. Three-quarters of an inch fell Tuesday night at the automated recording station at Bostetter.

Goose Creek reservoir will fill, reported the May SCS snow survey bulletin. The streamflow forecast had been estimated at 178 percent of average the first of April; it was now estimated to be 241 percent.

On Wednesday, May 2, state SCS official Rich Lavender went with J.J. McLaws to take snow readings at Badger Gulch. It was raining at the lower elevations and snowing up high. They found that the snow-moisture content at Badger Gulch, 9.6 inches in an average year, was 22.7 inches. Kent Foster recalculated the water budget. Getting rid of 8,000 acre feet would not be sufficient. A minimum of 12,000 acre feet would have to be released.

That same Wednesday Terry Bingham met Corps officials Bob Thuring and Tom Tate at the Burley airport for the first of many helicopter flights over the reservoir and drainage area. Because there was room in the helicopter for one more person, Terry invited Golden Valley farmer Roland Bean to come also.

Roland, whose farm was on the Goose Creek channel, had been concerned about flooding ever since Burke Scholer had talked with him. He had become more concerned the past weekend as he talked with people from Grouse Creek. "The water's backed up in our pasture," someone told him. "We've never had water there." The oldest residents had never seen so much water. "It's all coming your way," they warned Roland.

As the helicopter flew over the reservoir and the drainage area, Roland could see for himself what his Grouse Creek friends had described. The water was running high in Goose Creek and its tributaries. It was filthy water, full of trees, brush, all kinds of debris. Roland was horrified. In spite of everything he had heard from his Grouse Creek friends, he was not prepared for a sight like this. "I couldn't wait to get that helicopter on the ground so we could go after it," he said.

After flying over the drainage area, the helicopter flew toward Murtaugh Lake. The ground below was like a giant topography map. They could clearly see the old Cottonwood drain that went to the lake. Six miles from the drain was an abandoned canal that went from the Oakley West Canal northwest to Golden Valley.

Sending the reservoir water along that route to Murtaugh Lake seemed like the perfect solution to Terry and Roland.

Tom Tate did not think so. Tom had seen many floods in his life. The flight over the reservoir showed the flood potential, but it did not frighten him as it did Roland and Terry. Tom was concerned about the political problems that a diversion to Murtaugh Lake would bring. They would have to take water out of its natural channel and cross two drainage areas. They would need permission from the Twin Falls Canal Company, the Milner Low Lift Irrigation District, two highway districts, and all the farmers along the route. Tom had learned long ago that construction was easy compared to getting rights-of-way. Always there was someone who would not cooperate. Their best shot, he believed, was to find a flat place on BLM land somewhere, dump the water there, and let it percolate into the ground.

The helicopter landed in Oakley for an 11:00 A.M. canal company meeting. Secretary Klyda Wake remembered the stir the helicopter created in their little town. Although Klyda had heard people talking about Tom Tate, this was the first time she had seen him. She noticed that he walked like a man whose feet hurt, and wondered whether he had arthritis. Klyda was excited to see the helicopter and the officials. Nothing like that had happened in Oakley for years. She was not concerned about the reservoir; she could not imagine a flood happening in Oakley.

At the meeting Tom talked about taking the water down the West Canal into the desert, and if necessary going to Murtaugh Lake. They organized a committee to meet with the Twin Falls Canal Company. Roland Bean asked for permission to clear brush from the West Canal and widen the road along the top. The directors told him to go ahead.

"They gave me permission, but I think they were surprised when I moved so fast," said Roland. Immediately after the meeting, he headed for the Burley Irrigation District office. Roland had worked earlier with BID manager Evan Rasmussen and knew that BID had the small equipment needed to shape the canal.

Evan asked Leon Adams to help clear the brush so he could work with his excavator. "I had a tractor with a loader," Leon said. "I hooked onto the Russian olive trees to drag them away. Somebody was there with a saw to saw them down. We pulled up some fence posts and cut the wire, began clearing off the sagebrush and wild roses."

Roland went next to the SCS office to look at maps. Could the SCS begin surveying the canal? he asked Kent Foster. The banks needed to be smoothed and the road along the top of the canal needed to be widened so heavy equipment could get in. Burke Scholer had worked on those banks earlier so he knew what machinery was needed and what had to be done. Could the SCS get started right away?

"I was confused as to who would do what and who was responsible overall!" Kent noted in his log.

At three that afternoon Kent and Roland went to the law enforcement building for a meeting. County commissioners, Corps of Engineers personnel, canal company directors, state and local government officials, reporters, church leaders, farmers—everyone concerned about doing something to avert flooding—attended the meeting. The Corps officials proposed dumping water onto a flat place in the desert and creating an artificial lake. "Goose Lake", as the SIP dubbed it, would be about six miles northwest of Oakley on BLM land. It would be 1½ miles in diameter, with an average depth of five feet. The Corps would build a six-foot-high dike to keep waters from flooding surrounding areas.

It appeared to be the best plan. County officials would not have to obtain easements because they would not have to cross private property. Tom Tate estimated that it would cost about $200,000; the Corps would pay between 30 and 80 percent. The head of the Corps of Engineers in Washington, D.C., had to authorize the Corps' assistance, which Tom estimated would take a week. In the meantime, they could decide on the exact place for the lake and the local people could begin enlarging the West Canal.

As soon as the meeting was over Roland Bean headed for Oakley where he and others began tearing out the fences along the canal company right-of-way. "That made people a little upset," Roland said. "Those fences were holding cattle in. We had livestock running all over the country. If I had known they were holding cattle, I wouldn't have cut the fences, but I didn't think about that. All I thought about was all that water coming our way."

"Roland really wanted to get started," said County Commissioner John Adams. "He may have antagonized a few people, but he needed to do it to get them awake."

That same Wednesday afternoon the Oakley Stake youth groups and leaders met to fill sandbags. The Oakley Third Ward youth group, consisting of five adults and thirty-five to forty teenagers led by Beverly Franks and Kim Cranney, drove to the area east of Oakley by Birch Creek. Trucks had dumped sand in several piles. Their group was assigned was to fill fifty-pound potato sacks and stack them on pallets to be hauled where needed.

"The kids worked really good," Kim said. "Of course there was a little horseplay and a few sand and water fights. Somebody who came by and saw what we were doing said he didn't think it looked very well organized.

"'You ever tried to organize thirty or forty teenagers?' I asked him.

"It may have looked disorganized, but those kids worked hard. I was proud of them. I told them this was something that they should remember, something to tell their kids and grandkids. All my life I've grown up hearing about the bad winter of '49. I told them, 'Those who lived through the winter of '49 have nothing on us. From now on people will remember the spring of '84, and you are here and are part of it.'"

They had almost finished filling sandbags when someone called Kim on his two-way radio.

"The Birch Creek feeder canal broke," said a voice. "The water is headed your way. Get those kids out of there!"

"I didn't mean to scare them," Kim said. "I figured we had about an hour and a half before the water would get to us, though it came sooner than I thought. But you'd have thought it was the Indianapolis 500 the way everybody piled out of there."

The Birch Creek feeder canal, built when the dam was constructed, was designed to carry Birch Creek water to the reservoir. A headgate in the mountains above the dam regulated the flow from the creek to the feeder canal. When the canal washed out, a mass of water and mud came pouring down the old Birch Creek channel.

Pella youth and adults filling sandbags at Joe and Gwen Beck's farm, May 1984. These were among the many groups that filled sandbags throughout the crisis. Fred Hawker stacks bags while behind him Brandee Beck and Angie Beck form part of an assembly line. (Photo courtesy of Gwen Beck.)

Commissioner John Adams had gone with his teenage boys to help fill sandbags at Pickett's gravel pit. They had barely finished filling the 3,000 bags when news came of the broken feeder. John and others loaded and hauled the bags and headed for the problem area, near where Basin and Birch Creek roads came together. "It was a big stream," he said. "You didn't want to get in it. It could sweep you off your feet."

Garth Greenwell brought a backhoe to help dig a trench, Wayne Mullen remembered. "We were trying to trench from Tolman's house to the culvert by Bob Anderson's house. The hoe lost its footing— an arm of the hoe was out in the water. That and the front end bucket were the only two things touching. The footing was gone from underneath the tractor; the stabilizer arms were suspended. Everybody was scared to death about it, afraid it would tip sideways into the water. Mike Cranney said, 'Move out of the way, I'll get it off.' He got it off without tipping it over."

They diverted the water into the culvert, "then it was too much for the culvert," said Wayne. "We dug it out and let the water go. It went behind Bill Babbitt's, through his pasture—took out a lot of his topsoil—to Mildred Gorringe's and into the East Canal."

"They routed the Birch Creek water down through Mildred Gorringe's place because she didn't have anything planted," said Loreen Gorringe. "People were trying to keep it off their property—running it onto someone else's. They ran it into the East Canal—had as much water as the canal would hold. They ran the risk of having a breakthrough. Nobody wants water when a flood comes."

People were concerned about LaVonne Pressley, who lived alone, remembered John Adams. Her house was surrounded by water. She could not get out, and no one could get in. No one could figure out how to reach her. "She'll be all right until morning," someone finally said.

LaVonne's first indication of trouble came when she looked out her window and saw that the culvert in front of her place was flooded. "I walked out and got stuck in the mud," she said. "I had been injured earlier and was using a cane. I tried to go out the back way, but the sub water was so high I couldn't make it. Then just as I was calling someone, my phone went out." From her window she could see people sandbagging to protect her garage.

"I was talking to Stormy Anderson when my phone went dead. She called the phone company immediately and told them that I had to have a phone, that I was surrounded by water. They had it fixed by the next morning."

The next morning "Ann Anderson and her husband brought water through the field to me. I didn't have water. I was surrounded by it, but didn't have any. They put a board across so I could walk out. They built a bridge so I could get the car out, but I didn't dare go back in. I left the car on the other side of the road. Stormy took me to get groceries. I had a wheelbarrow to carry stuff across."

When the Birch Creek feeder broke, David Babbitt, city of Oakley public works director, was in Twin Falls purchasing a 30-inch culvert. "I wasn't planning on doing anything about Birch Creek," he later said. "I figured the canal company and the Birch

Creek water users would take care of it. Oakley Council Chairman Maurice Matthews sent me to get the pipe to place under Oakley's Main Street. He thought a 30-inch culvert would be more than enough to take care of any problems."

The next day David exchanged the 30-inch pipe for two 48-inch culverts which he placed side by side under the road. "We had to put the culverts in with the water running," he said. "It took several pieces of equipment to hold them down and push them in. It was hard to get it done. The water was washing so much gravel in so quickly."

For a day, until the Birch Creek feeder was repaired, the water ran along the side of Main Street. "There were big washes all along the side of the road," David said. "It hit Manning's driveway and went shooting four feet in the air."

Wayne Mullen and other church emergency directors monitored Birch Creek above Oakley. "We checked it every night," Wayne said. "We would watch the water through the sagebrush ground to see if it had risen—and it had risen. There were no roads. We'd drive up there through the brush. We didn't dare get too close—it was hard to see where the water was spreading."

Water from Basin Creek, which had never before reached Oakley, spilled into some abandoned canals and onto canal director Daren Critchfield's property.

"There wasn't anything we could do about it," Daren said. "We could only have made ditches and diverted it onto someone else, and we couldn't do that. Everybody was complaining about the canal company, and we were the ones that got dumped on. The water cut channels through our fields and ditches, ran through our circle pivots for two months."

With so much water from Birch Creek flowing into the East Canal, farmers near the canal had difficulty keeping it from flooding their fields, Suzanne Critchfield remembered. "Every ditch everywhere in Oakley that would hold water was filled to capacity plus," said Wayne Mullen.

People had expected some flooding from Birch Creek, but no one had thought it would be so bad. As the Tolmans cleaned out their basement and LaVonne Pressley hauled her groceries in a wheelbarrow, everyone had the same idea. If that much water was coming from Birch Creek, what would Goose Creek be like?

Thursday at 7:00 A.M. Kent Foster met Mike Cranney and Roland Bean at the West Canal. "We felt it would be best to beef up the east bank and raise the water level about 2 feet in the canal and send all the water we can down to the north diversion point past W.B.'s feedlot," Kent wrote in his field notes. "Looked at the situation with the canal board and had a short meeting with the group and Corps at the canal office. Took shots at all thru crossings to determine flow restrictions of culverts, etc."

Bureau of Land Management officials made a quick assessment of environmental damage. Under normal conditions they would have spent a couple of months making environmental impact statements. Instead they made a rough survey while archeologist Isaac Barner checked the area for native American sites.

The Oakley Stake presidency called for volunteers. "We asked the bishops, 'Can you furnish such and such amount of trucks?'" remembered counselor Dallan Elquist. "The bishops contacted the people and asked them to help. Once we called the bishops, we could forget it. We knew the trucks would be there."

People came who had work to do on their own farms. People from View and Unity helped, though they were not in danger of being flooded. Many volunteered without being asked.

Although he had retired from farming because of his health, Harold Bowers worked on the canal. "Harold was the disaster person for our ward," said his wife Lila. "He was not feeling good, but he took the truck to Oakley and worked many days."

"My farm was having its own little flood," Dallan Elquist said. "I was leaving floods at home to help at Oakley. The flood here did not threaten my home, only the fields and ditches. Our man could handle it."

"May 3, 1984, Thursday," wrote Delma Bean, Roland's mother, in her journal. "Roland and Mike Cranney are trying to get people to bring heavy equipment in to build up the canal that comes out of the Oakley dam because there's going to be a flood if they don't get something done. Bruce took the backhoe and Allen the patrol, and they took the carryall and are working to clean out the canal and build the banks."

Bud Hoffbuhr, owner of Hoffbuhr Excavating Company, began enlarging the canal at Whiteley's feedlot. "Roland Bean got me involved," Bud said. "He took full responsibility. 'If worst comes to worst, I'll pay you myself,' he told me. 'Something has to be done, and done right away.'"

Bud attended some of the canal meetings. "I enjoyed the meetings," he said. "I enjoyed seeing people pulling together and being neighborly. They forgot all their petty differences."

The Relief Society, the women's organization of the LDS church, provided lunches for the crews. "President Gerratt stressed that he wanted good lunches," said Stake Relief Society President Suzanne Critchfield. "He wanted more than a couple of pieces of bread slapped together with some meat. He told us what he wanted on the sandwiches—lettuce and cheese slices, a variety of meats. We added cookies, a candy bar, a big apple or some other fruit, all wrapped in individual sacks." The men distributed the lunches, plus a can of soda pop.

"At first my counselors and I put the lunches together, along with anyone else who wanted to volunteer," Suzanne said. "Bill Garrard (Jer ard′) was High Council advisor to the Relief Society. He volunteered to help and sent his wife."

As the need for lunches increased, Suzanne asked ward Relief Societies for help. They rotated through the wards, with each responsible for lunches on a certain day. Usually about ten women met at the church where they formed an assembly line.

"The ladies were glad to do it," Suzanne said. "Everybody wanted to help."

Canal Company secretary Klyda Wake remembered that she was making sandwiches one day when a neighbor, not a member of the church, came in and asked, "Can't I help too?"

"We put her to work," Klyda remembered.

As volunteers worked on the canal, Terry Bingham, the county commissioners, and attorneys completed the paper work needed for the Corps of Engineers to participate in "Advance Measures aimed at reducing or preventing flood damages." They estimated how many acre feet of water would spill over the reservoir and what the property damage would be. They agreed to provide all land, easements, and rights-of-way, and promised to hold the federal government free from any damages. The governor of Idaho had to declare a state of emergency. Paul Massie said he would handcarry the county's request to Governor John Evans the same afternoon.

Since becoming governor in 1977, John Evans had dealt with drought, volcanic ash from Mount St. Helen's eruption in Washington, and the collapse of Idaho's economy. Now he faced disastrous floods along with continuing recession. Paul Massie arrived in Boise late Thursday with Cassia County's request for disaster assistance. Friday morning the state officials and lawyers were hard at work drafting the emergency declaration.

But even as state officials drafted the declaration, those on the scene in Oakley were having second thoughts about Goose Lake. There was no large, flat place to put the water. Kent Foster, an engineer, had made some calculations. If they built on a slope, they would need a mammoth structure which would be very expensive and possibly not safe. If the dikes broke or the soil did not soak up as much water as anticipated, there would still be flooding.

A canal to Murtaugh Lake would cut through private property, but it would handle the water. In spite of possible easement problems, it seemed a better plan. Tom Tate and Kent discussed the situation Thursday night, then Tom called Corps' headquarters in Walla Walla. "No more Goose Lake," wrote Terry Bingham in his notes the next day.

Meanwhile canal director Daren Critchfield was thinking of another alternative. "I remembered my father saying that because the reservoir was narrow at the bottom and wide at the top, the top few feet would hold almost as much water as the rest of the reservoir," Daren said.

He went to the reservoir and measured the distance from the top of the spillway to the top of the dam. It was twelve feet. If the spillway were raised by only a few feet, the reservoir could easily hold the extra 8,000 to 12,000 acre feet of water. Daren talked to the other canal directors. "Let's petition the Idaho Department of Water Resources for permission to raise the spillway three feet," he suggested. It would be a lot cheaper than building a canal, and would keep the water in the reservoir where they could use it.

Part of the reservoir spillway, 1971. (Photo courtesy of Merlin Stock.)

Winslow Whiteley was skeptical, but agreed to go along if the Department of Water Resources gave permission.

The Department of Water Resources would not consider raising the spillway, the officials told the canal company directors. That might place in jeopardy the integrity of the dam.

By Friday, May 4, several volunteers were working on the canal. The city of Burley had sent water trucks, dump trucks, and compactors, noted Terry Bingham; the Burley Highway District sent sheep foot compactors and trucks; LDS church volunteers furnished dump trucks, loaders, and carryalls, plus lunches for everyone.

The big loader belonging to the Burley Highway District was at Heglar, in the eastern end of the county. Jones Leonard said they could use it if Terry could find a truck to transport it. It was in a pit east of Kendall Jones' house, where Fuzz Wren used to live. The key was behind the windshield washer motor or above the door. Check the oil—use series three.

Terry Bingham also noted, "Call sheriff and get an extra patrol in the Oakley area. Need a road block.

"!!Keep tourists out!"

West Canal construction, with Middle Mountain in the background. Farmers donated their time, equipment, and crews; construction companies worked without signing contracts. (Photo courtesy of Wayne Mullen.)

That Friday Terry went with Weldon Beck, Winslow Whiteley, Roland Bean, Mike Cranney, and Al Barrus to meet with the Twin Falls Canal Company. Like the Oakley Canal Company, the Twin Falls Canal Company was privately owned; the farmers who used the water were the stockholders. Their reservoir, Milner Dam on the Snake River, was one of the first and most successful of the Carey Act projects. Water flowed from Milner Dam to Murtaugh Lake, which originally had been a dry, low area, then into irrigation canals.

Maurice Klaas, a local farmer and agricultural real estate broker, was chairman of the board of directors. He arranged for the Oakley people to meet with the Twin Falls Canal Company lawyer, manager, and board of directors.

"Everyone is jealous of their own irrigation system," Maurice later said, "but we did not think that taking the Oakley water would do us any damage." They were concerned about starting a precedent and about liability if the flood damaged their system. They wanted clean water. Some of their structures were old and might be damaged if debris hit them. But overriding these concerns was their desire to help the people of Oakley keep their valley from being flooded.

"We felt good about it, felt good from day one," Maurice said. "Cooperation is important. You never know when you're going to be in a situation where you need help from someone. There's enough misery in this world without people not cooperating with their neighbors."

The Oakley people could go ahead with their plans, Maurice told the committee. Any problems could be ironed out later.

"Mr. Whiteley reported on their meeting with the Twin Falls Canal Company," recorded Secretary Klyda Wake later that day. "They were very agreeable to the plan of taking the water down the old Cottonwood canal into Murtaugh lake and assured the committee of their cooperation. . . .

"Bud Robinson was appointed as supervisor of the work being done on the west canal. . . . Had been planning on raising the west canal to hold 20,500 [miners] inches. It was suggested that it be raised to hold 35 to 40,000 inches [700-800 cfs]. Soil Conservation men left to start raising their stakes for more capacity.

"Liability insurance—The premium is due and Jay Robinson moved that attorney Bywater see if the company can get an umbrella of 10 million dollars."

Immediately after the meeting, attorney Steve Bywater began working on easements. He needed the names of all landowners along the route and legal descriptions of their properties. "I didn't sleep at all that night," he said. "I spent a day and a night looking up property titles and preparing the documents for signatures."

Saturday, May 5, the work on the canal began "in earnest" wrote Oakley stake historical clerk Donald Clark. Local construction companies—Kloepfer Concrete, Pee Wee Construction, Lynn Watterson, Gordon Paving, Hoffbuhrs, Thorntons—farmers from Oakley Stake, crews from the city of Burley and Burley Highway District built up the bank on the low side of the West Canal. Cats dozed dirt from the mountain while loaders dumped it into trucks that carried it across the canal. They planned to raise the bank five feet.

Lynn Watterson in a grader pushing up the bank on the canal. (Photo courtesy of Paula Watterson.)

"I just scooped up bucketloads of dirt and dumped," said Kloepfer loader operator Lawrence Burch. "Trucks were always there waiting for me. I had to be careful not to knock the bottom out of the grain-bed trucks. Kloepfer truck beds were made of thick steel, but the farm trucks were tin."

Having so many volunteers did not cause a problem, he remembered. "Everybody tried to work together using good common sense, and stay out of each other's way. The only problem was getting the trucks from one side of the canal to the other. There was only one bridge that both empty and loaded trucks had to cross."

None of the construction companies had signed contracts. They were not sure who, if anybody, would pay.

"I wasn't worried," said Ernie Straubhaar (Stru′ bar), vice president of Kloepfer Concrete. "I trusted Winslow and the others. I figured at worst we'd get our money late."

"Nobody worried about pay when it became an emergency," said Bud Hoffbuhr. "I just hoped I could get enough to cover expenses and meet my payroll."

Larry Adams, owner of Adams Petroleum, arrived with his diesel truck. "Guys, you've got to have diesel," he told the canal directors. "We'll figure out details for payment later. Just tell me how much you need to keep running."

"We started with one truck on Saturday afternoon," Larry remembered. "We just dumped it in. We donated petroleum for all volunteers. We said, 'You're donating, so we're donating.' Then one of these guys for hire would hear that there was free gas and would come running up. We'd say, 'No deal. You charge, we charge.'" Larry planned to donate four thousand gallons of fuel, but donated twice that amount.

Many volunteers furnished their own gas. "Wade Baker opened up his tanks for his crew and anybody else, and filled all comers," Grant Wyatt said.

Construction crews and volunteers worked late Saturday night; then all activity ceased. Sunday was the Sabbath, the day of rest. Some people felt that because it was an emergency they should continue working. Didn't it say in the Bible that if an ox fell into the mire on the Sabbath day it was all right to pull it out?

This was not that much of an emergency, others contended, and they needed the Lord's blessings for a successful completion of the project. The ox was not in the mire yet.

For LDS church members the first Sunday of every month is "Fast Sunday," a day in which they go without eating for twenty-four hours and donate the money saved for care of the poor. On May 6 the bishops of the region told their congregations that this was a special fast day during which they should pray concerning numerous weather-related problems everywhere. They were to ask the Lord to temper the elements so that severe flooding could be avoided.

This Sunday was special in a different way for the Oakley Stake leaders. Elder David B. Haight, general authority and member of the church's Council of Twelve Apostles, was coming from church headquarters in Salt Lake City to dedicate the new Grouse Creek chapel. Although he had spent most of his adult years in California, Elder Haight had been reared in Oakley. People in Cassia County felt that he was "their" general authority.

For the Oakley leaders the trip to Grouse Creek was sobering. Water was everywhere, in places it had never been before. It was one thing to hear Grouse Creek residents tell about it, but quite another to see it. The threat of flood hung over the proceedings of the day.

Elder Haight offered the dedicatory prayer. Near the end of his prayer he asked that the elements be tempered, adding, "We pray, Heavenly Father, that thou wilt hear our pleadings and hear our prayers, that the weather might be steady in such a way that the snow will not melt too fast, and the water will be able to be taken care of."

Oakley Stake President Rex Gerratt believed that the prayer was a blessing and a promise and that the Lord would help them. They would have to work hard, but they would have the time they would need to get the job done.

In Twin Falls that same Sunday afternoon, Roy Beams, manager of Armco Construction Products, had come to his office to catch up on paperwork. He had barely started working when the telephone rang. It was Ray Kimber, manager of the Oakley Highway District. They were going to build a canal to Murtaugh and needed pipe right away, Ray told him. Carl Gooch from the Murtaugh Highway District had suggested he call Roy because Roy would deliver pipe any time. How much 36- and 48-inch pipe did he have on hand, and how quickly could he get it to Oakley?

He'd load up a truck and leave immediately, Roy said. He'd bring him everything he had and would order more from Salt Lake. The highway district did not have the money to pay for everything right away, Ray told him.

"No problem," said Roy. "Let me know what you need and I'll keep you supplied. We'll worry about finances later."

That same afternoon Kent Foster, Tom Tate, and Terry Bingham drove through the sagebrush searching for the best route for the diversion canal. They decided to follow an abandoned canal across BLM ground to the Golden Valley area, go along a gravel road for four miles, jog northeast into a couple of gravel pits, follow the Murtaugh 2900 North road for a half mile, then along the natural drainage to Murtaugh Lake. The canal, including the enlarged West Canal and the natural drain, would be twenty-three miles long.

They believed that the gravel pits, which were about ten feet deep and covered eighty acres total area, would absorb a lot of water. With good soil percolation, the gravel pits might absorb almost all the water, leaving only a little to run to Murtaugh Lake.

The road they chose was the same road Bill Cranney had selected for diverting Cottonwood Creek water. It had a steep grade, Kent noted, which could cause the water to flow fast. But any other route would cut through the middle of fields causing damage to crops and disrupting irrigation systems. By tearing up the road, they would minimize damage to farm land.

They could slow the water by building a wide, shallow canal which would allow the water to spread and percolate into the ground. A wide canal was risky because water could run from side to side and erode the banks. But the canal would not have to carry a lot of water—only about 250 cfs, plus the runoff from Cottonwood. They could compensate for Cottonwood water by decreasing the reservoir flow. A canal the size they envisioned could easily handle up to 400 cfs if necessary. It was the best route, they decided, the route that would most likely avoid flooding and minimize damage.

Because there was no time for detailed surveying, they would have to work mainly from quad maps. Kent Foster knew he was not following approved SCS guidelines. But he also knew that if they followed established procedures, the planning alone would take 2½ years.

"We used our professional judgement to do the best we could as quick as we could," Kent said.

They had found their route. Now all they had to do was construct the canal before the water came.

Widening the West Canal.
(Photo courtesy of Larry Adams.)

Chapter Six

May 7–May 12

"Met with County Commissioners, Tate, Bingham, etc., and discussed route," wrote Kent Foster in his field notes for Monday, May 7. "Met at law enforcement building to discuss other essentials such as reducing flow and work on the West Canal to get done quicker."

Time was running out. It had rained again on Saturday; the National Weather Service was predicting higher temperatures. If it turned hot, the water would come down all at once.

Terry Bingham described the Corps' requirements for participation: a government unit as a sponsor to supply the rights-of-way, a legal agreement holding the U.S. Government free from liability, a local agency to relocate fence, power, and road crossings, and to operate and maintain the canal.

"Have the county attorney sign an affidavit that the three county commissioners have the right to sign," noted Terry.

Mike Cranney called Idaho Senator James McClure, asking him to stress to Corps officials in Washington, D.C., the necessity for quick action. Tom Tate had talked to contractors. They were ready to move their equipment as soon as he gave the word.

"Tom Tate was more knowledgeable about heavy equipment than anyone I had ever met," said Terry Bingham. "He knew every contractor in the region, and the types of equipment they could furnish. He had their phone numbers in his head."

The Corps could not begin until all easements were signed, Tom warned. He did not see how they could get easements in less than a week.

Another bridge was needed to eliminate the bottleneck on the West Canal. "Someone tried to get flat cars out of Utah because those had been used as bridges before," Kent Foster said. "They couldn't find any. Somebody mentioned the National Guard bridges. Others said we would never get those. The National Guard didn't let them go."

Terry called Adjutant General James Brooks. The county could use a bridge if the site were suitable, General Brooks told Terry. He would send a man immediately to check on it.

Terry and the commissioners went to Oakley for another meeting with the canal directors. "Financial matters were considered this morning," wrote secretary Klyda Wake. "It was Mr. Beck's opinion that the first responsibility for finance would be to the Oakley Canal Co. and the County will back up. Mr. Beck and Mr. Whiteley are to go to the bank to see about getting financing."

In the middle of the meeting Maurice Klaas, Clarence Hollifield, and Warren Travis from the Twin Falls Canal Company arrived. They had driven through the sagebrush following the route of the canal.

"When we saw the route, we did not see how it could do us any damage," Maurice said. "We thought the water would never get to Murtaugh Lake. We thought it would be dissipated into the gravel pits. One thing that scared me was seeing that dam—knowing the devastation that would occur if the dam went or they didn't get rid of the water—knowing the job it was to get rid of water—seeing that reservoir as full as it was with the water still coming."

The canal directors asked Ernie Straubhaar, vice-president and manager of Kloepfer Concrete, to supervise construction of the West Canal. No contracts were signed nor terms discussed, Ernie later remembered. "Winslow just said, 'Run the project and get her done.'"

"We said, 'Ernie do it. We'll talk about it later," remembered Winslow.

At the time Ernie did not think it would be a big job. "It was routine," he said, "a matter of raising banks. At first we worked only in the daylight. I didn't think it was that big an emergency."

Al Barrus, Ernie Straubhaar, and Terry Bingham. (Photo courtesy of J Malan Heslop, Church News.*)*

Part of the West Canal wound through the Whiteley cattle feedlot. To enlarge that part, the workers dug a channel next to the canal, leaving a ribbon of dirt about eight feet wide between the canal and channel. When everything was complete, they would cut through the ribbon. In the meantime, Jay could run water in the West Canal.

"Ernie oversaw everything," Commissioner John Adams said. "It was logical and smart. Everything they did was the right and the easiest way. Farmers are kind of hard-headed, but they worked together beautifully." He watched the small farm trucks packing and unloading dirt. From a distance they looked like a bunch of ants.

"We had a pilot man to guide the farm trucks," said stake disaster director Wayne Mullen. "We coordinated with Ernie Straubhaar to find where he wanted the dirt. Someone was standing there to direct the truck. He'd say, 'You're back far enough—dump.'"

Clem Stanger helped spread the dirt. "There were so many people with vehicles," he said. "Every little farmer in this valley was out there. They didn't have huge equipment like the Corps later used. There were tractors with scraper blades, ten wheelers, one-ton trucks."

Farm trucks building up the banks of the Oakley West Canal. The foreground shows dirt that will be removed to widen the canal. Note the ribbon of earth between that dirt and the canal. (Photo courtesy of Larry Adams.)

J.J. McLaws and other SCS employees surveyed the route. J.J. was amazed that everyone worked together so well, especially when so many were inexperienced volunteers. If there was the "tiniest little problem," it was corrected within minutes. "It run so smooth it was unreal," he said.

But Don Holmes, owner of Maverick Construction, thought there was a lot of confusion. "Some guy would push a load of dirt over there, and somebody else would push it back. It was not too well organized."

More volunteers arrived—crews from the Minidoka Highway District, farmers from Paul and Rupert stakes. Greenline Sales brought a new John Deere dozer for the volunteers to drive. Case Equipment brought a backhoe. "I asked the Case Power dealership if they could help us with a backhoe and they sent us a brand new backhoe," Leland Baker told the *South Idaho Press*. "You can't ask for more."

Greenline Sales hauled equipment for people, remembered salesman Gale Moncur. "People would call and say, 'We'd like to help, but can't get our loader or scraper to it.' We had a semi with a flatbed trailer. We'd haul it out there for them."

"I had bought a new tractor in March," said Oakley farmer Randy Hardy. "Someone came looking for a tractor to pull a crow's-feet packer, so I told them mine was in the yard. The next time I saw that new tractor, it was on a narrow, soft bank. I had no idea who was running it or if it would be tipped over. A lot of things went on like that with the equipment. We had to trust each other."

Oakley Highway District crews installed culverts as the roads were cut. "We first put pipes in at Whiteley's feedlot," said manager Ray Kimber. "We wanted to save a main route in and out of here. Then we started on the crossroads. As they was digging, we was following along behind, putting in pipes."

Ray also had crews repairing the Basin road and other roads that the creeks had damaged. The Birch Creek feeder had broken again during the weekend, pouring more water into Oakley.

"They had put more water into the feeder when it broke the second time," said Oakley public works director David Babbitt. "So much water was coming down Birch Creek. It runs the quickest of any creek around. It's fed by high snow and low snow. It kept running strong."

The farmers along Birch Creek were putting water in ponds, irrigating with it, "using water like crazy." The old Birch Creek channel ran through Oakley. David diverted some water into a ditch along the old channel and into the town irrigation ditches. The rest of the water, about 7,000 inches, ran along Main Street for a third of a mile, into another ditch and then into the East Canal.

"A lot of debris was coming down," said David. "I patrolled night and day keeping the culverts clean. The biggest challenge was at the dividing point. It was a high point—it kept silting in, kept graveling in. I'd have to take my backhoe right in the water to keep it cleaned out."

Tuesday, May 8, at 3:18 P.M., Corps headquarters in Washington, D.C., granted permission for the Corps to participate in "advance measures to avert a flood." The Corps would construct fourteen miles of the proposed twenty-three-mile canal. The local construction crews and volunteers, whom Terry Bingham had dubbed the "rag-tag army", would be responsible for the first five miles, which included enlarging four miles of the West Canal, and for making improvements along the last four miles of natural drainage leading to Murtaugh Lake. The Corps would begin work as soon as all easements were signed.

They needed easements from the Oakley and Murtaugh highway districts, the Twin Falls and Milner Low Lift canal companies, and all landowners along the route.

Oakley Highway District commissioner Bill Cranney met Tuesday at noon with the other commissioners. They signed an agreement in which the canal company accepted all liability.

Bill then met with the Murtaugh Highway commissioners who gave permission for the canal company to use a half-mile of their 2900 North road. "They said we could do it," remembered Bill, "but if it got that far with 500 [miners] inches of water they'd be surprised."

Everything was agreed to verbally, manager Carl Gooch remembered, and later put in writing. His crews were working before Wednesday, May 9, when the formal agreement was signed.

Tuesday, May 8, the Twin Falls Canal Company directors met with their two attorneys to discuss their agreement with the Oakley Canal Company. The Oakley Canal Company agreed to give twenty-four-hours' notice of water coming to the lake and would be responsible for any damage done. "The maximum flow rate, including flows naturally in Cottonwood Creek, shall not exceed 400 cfs, measured at the point where Cottonwood Creek enters Murtaugh Lake," the agreement stated. Although they had to cover all possibilities, Maurice believed it was only a formality. He expected very little water to enter Murtaugh Lake.

Tuesday at 4:00 P.M. a delegation from Oakley met with the board of directors of the Milner Low Lift Irrigation District. They needed permission to take the flood waters across the Milner canal. "We were expecting them to come," said manager Jim Bright. "We'd heard about their problem through the media. We didn't want their water to intermingle with our water, that was our only concern."

The gravel pits would soak up most of the water, the men thought, and the pits were between the reservoir and the Milner canal. They talked about taking the flood waters under the canal in a two-foot pipe, but decided to use a five-foot arch siphon instead. A siphon that size would easily handle 250 cfs, the maximum flow that was expected to reach the canal. The Oakley Canal Company would pay all costs and assume all liability. By six that night the agreement was signed and notarized.

Meanwhile Terry Bingham, Steve Bywater, Roland Bean, and Mike Cranney had talked with the landowners. "Asking for permissions was not an easy job," Roland said, "but the people were willing. Those people lived clear over in Murtaugh, in another county, yet they were willing to let water come their way."

"Most people were glad to help out," remembered Mike. "Most said, 'We'll be a neighbor. We know it's an emergency.' We got lots of comments like these."

"We had no hesitation in signing a right-of-way," said Golden Valley farmer Carl Van Tassell. "We knew what was going to happen. We knew what a lot of water would do if it got loose."

"I gave my permission," said Murtaugh farmer Melvin Tilley. "I figured we really could help people by letting the water come."

"Sixty to 70 percent of the people were very cooperative," said attorney Steve Bywater. "They had a few questions, but once we explained why and how, they were willing to sign. We talked to them about fifteen minutes. Others were less willing. Some had more to lose. We would cut through their ditches and roads. Some wanted to contact their own attorneys."

Peggy Kelly, who owned land near Murtaugh Lake and taught school at Murtaugh, was reluctant to sign an easement, her daughter Amber McFarland remembered. "Mike Cranney and another man came to her in the evening and needed an easement," said Amber. "She felt the guys were picking on her because she was a woman, alone. They came late at night, unannounced, when she had to be teaching the next morning. They were not very diplomatic. The guys were telling her she needed to sign, that they would lose their ground otherwise. They were going to push it through no matter what. She and Mike had been friends. They're friends again, but for awhile it was a sore spot with her. She wanted certain things in the agreement. They had to rewrite it two or three times before she would sign."

Idaho industrialist J.R. Simplot owned property along the route; his manager refused to grant permission. Simplot would have to sign the easement, the manager said, and he was vacationing in Hawaii.

"I'll take care of that," said Winslow Whiteley when he heard about the difficulty.

"Jack and I had been friends since 1930," Winslow later said. "I knew where to get in touch with him. I gave him a call and explained the situation. 'You want to come down here and look it over?' I asked him.

"'I don't need to come down,' he said. 'If you say you need it, take it. Sign anything for me that needs to be signed.'"

Tuesday night, May 8, the last of the private landowners signed easements. "Even those who had some reservations about the project were public-spirited enough to go along," said attorney Steve Bywater. "If we'd have had to go to court, it would have been impossible."

In all the years that he had worked with the Corps, Tom Tate had never seen easements obtained so quickly. "I've never seen the degree of cooperation we have here," he told the *South Idaho Press*. "It's been fantastic. I've been doing this for 30 years and I've never had it this way. There's nobody crying about nothin'. Everybody just says, 'Anything you want to do, go do it. . . .' Usually with something like this there's always somebody fighting you. But there hasn't been anybody here who's not 100 percent behind it. . . . It's been like that since I got here. Everybody just says do it and we'll worry about the technicalities later."

Corps-hired equipment began moving to Burley Tuesday night. The equipment was huge, people remembered—some D–8 and D–9 cats that needed an escort as they were trucked along the highway. Corps officials Donna Street, Larry Johnson, and Larry Pfefferele arrived from Walla Walla to supervise. By Wednesday afternoon they were digging the canal.

"I couldn't believe all the equipment that was working," said Carl Van Tassell. "Within two miles there were at least a dozen cats."

They worked day and night, running two 11-hour shifts, with an hour between each shift for maintenance and operator change. Tom estimated that they would finish the canal in a week.

Bud Hoffbuhr, who had not been sure who was going to pay him, found himself working for the Corps. "I liked working for them," he said. "They were so well organized. They were not like regular bureaucrats." As one thing was finished, they'd have the next project all lined up. There was no time that operators sat around waiting for someone to tell them what to do.

Tom Tate and Donna Street from the Corps of Engineers supervising the middle section of the diversion canal. (Photo courtesy of the Walla Walla District, U.S. Army Corps of Engineers.)

"Tom Tate was the smartest man I've ever had the pleasure of being acquainted with," said Bud. "He knew construction, he knew organization. Emergencies were his business. He'd been doing it for years. He had trouble walking, but it didn't slow him down any. He was the most impressive guy I'd ever been around."

Bud needed more workers to keep his equipment running twenty-four hours a day. A couple of local men who worked in gold mines in Elko, Nevada, knew heavy equipment; Bud hired them. Paul Truxal, whose wife owned Adam's Cafe, had previously worked in construction; Bud hired him. Paul cooked all day and ran a backhoe at night. When word of the project reached the press, people from other places came looking for work, remembered Bud, "but I tried to stick with people I knew." Bud organized his crew during the day, and ran a backhoe at night.

"We were running around like wild men," he said. "We had the only excavating equipment in the area that could work in all that wet stuff. It was hard for me to go to sleep at night thinking about all that stuff out there running and me not being around to see it."

Carl Van Tassell remembered how considerate the Corps officials were. When the equipment operators first began digging near his place, they pushed a lot of dirt onto his field. "The head of the Corps of Engineers came by," said Carl. "I talked to him about it and asked if he couldn't get somebody to supervise, so they wouldn't push it onto my field so far. Up until then they'd been digging through desert land, hadn't been through anybody's field. He ran a line out here and said they couldn't go beyond it. They narrowed the channel a little and deepened it. They were real courteous."

Volunteers from Rexburg arrived. The Cassia farmers had helped them after the Teton Dam disaster; they had come to return the favor. "They came out of pure brotherly love," said Commissioner Weldon Beck. "All they asked from us was a place where they could sleep and shower."

Milo Siepert, a partner in the Edstrom Construction Company at Rexburg, was one of the volunteers. He brought three of his construction workers plus loaders and dump trucks. The LDS church in Rexburg had asked for volunteers, Milo remembered. He and his partner Delbert Edstrom sent equipment and a crew. "We talked it over and decided we would not charge the Oakley people for anything," said Milo. "We were in the Teton deal. We lost our home, trucks, shops—everything. I had my pickup, my wife drove our car out—that was all we had. I don't know what we'd have done in our valley if people had not helped us."

Milo took his loader to the canal near Oakley and began moving dirt. "It was not much different from what we usually do, except for the hours," he said. "Usually we'd work about ten hours, but on that project we was punching all the hours we could. We'd take care of the maintenance on our equipment at night. The people there fueled us and fed us and took care of us. I didn't know anybody in the area, but all of them was good and friendly. I got to know a lot of faces—never did know names."

The National Guard brought an armored vehicle launching bridge which they installed near the dam. In two sections on a hinge like a pair of scissors, it was lifted into the air, opened up, and lowered to the ground.

Although the bridge helped to speed construction, "it was not a lot of fun to drive across," said dam safety official Loren Holmes. "It was a steep access road to the bridge. You'd have to get lined up exactly right. Once you started up the road, you couldn't see the bridge over the top of your hood. The bridge had two tracks, with

The National Guard armored vehicle launching bridge being installed at Whiteley's cattle feedlot. (Photo courtesy of Duff Woodhouse.)

a space in the middle. You had to be right on track. There was a little rim on either side, but no guard rails."

"It was made for tanks and vehicles with tracks," said dam safety manager David Hollingshead. "It peaked abruptly. When vehicles tried to cross, they'd lose sight reference. Sometimes a wheel would drop in the middle of the bridge. Then they'd need a tow truck, or they'd jack up a wheel. That would snarl traffic."

When Dave was at the bridge, he tried to guide traffic across. "We were measuring flow at the bridge," he said. "It was a little hazardous. It was best to direct traffic."

Allen and Loren drove from Twin Falls almost every day to measure the water being released in the canals. "We measured the flows every time they changed gates," said Loren. "The canal company had a gate rule curve which showed that the gate at a certain height would release a certain amount of water, but they did not know how accurate it was for high volume. They had never released so much water before."

As Kim Cranney watched all the activity, he remembered thinking that "it was the best working together of local, state, federal

A pickup in difficulty on the bridge. "The tire of the pickup hit the rim," said Allen Merritt. "The tire blew and physically lifted the pickup and pushed it partially off the bridge." (Photo courtesy of the Idaho Department of Water Resources.)

government, and church that you would ever see. The right people were in office. Tom Tate was a good man—we were lucky we had him. He was a common guy to be the head man, not your prima donna bureaucrat."

"Tom Tate was good," said Bill Cranney. "He listened to local people and would blend in with them."

Although Terry Bingham was technically in charge of the local construction crews and volunteers, he did not try to boss the work on the canal. "Ernie had the experience in construction, in handling men and equipment," Terry said. "I would not have presumed to tell Ernie what to do."

"Ernie Straubhaar run day and night," said Winslow Whiteley. "He did a good job. He was very instrumental in the emergency. At the tip of the emergency, he was totally in charge."

Along the Cottonwood drainage at the Murtaugh end of the canal, farmer Leland Baker supervised the work. "I didn't do anything special to inherit the job," he later told the *South Idaho*

Press. "I was just helping out like everyone else around here. The job just sort of evolved."

As president of the Mini-Cassia Search and Rescue Unit, Leland knew how to organize volunteers. His pickup with its multiple radio frequencies, telephone, portable gas tanks, and "Baker's Bakers" slogan became a mobile base of operations.

Roland Bean worked with manager Jim Bright on the Milner canal. Jim turned the water out of the canal for a day so they could install the arch siphon which would carry the reservoir water. Melvin Tilley, whose farm was next to the Milner canal, was amazed to see what they were doing. "I said, 'Good golly, what do you need that big a pipe for? You'll never get that much water.'"

The Oakley people brought "cars, trucks, backhoes, everything," to protect his place, Melvin remembered. "They built the road up with dirt to keep water from going over. We had to get to a pasture. They brought a cat, come in and put culverts in. All of them worked together and fixed it up real good so we could get across."

"Roland Bean smoothed out people's ground, did a lot of things to help people feel good about the canal coming through," said Rex Gerratt. "He made sure that the equipment was driving in certain areas and not on people's crops." With machinery and crews from the family business, "Roland was putting everything he had into it."

"Dad and Bruce were running the farm," Roland said. "I was one hundred percent on the canal. I'd get up at five and be on the road—not home until well after midnight. Then I couldn't sleep—thinking about all the things that needed to be done. But we were able to do things that normally we couldn't do. We had strength and stamina that we would not have had under ordinary conditions."

"People did things they never could have done—I was going to say in peacetime," said Kim Cranney. "And I guess that's right. It really was a war."

"I was amazed—was amazed—was amazed at how people pulled together and worked together to get things done," said Milner canal manager Jim Bright. "The Oakley people were so good to work with. Local farmers had their crews out working We had a heck of a crew. The church brought lunches—good lunches. I'd never seen anything like it. I've never ever seen so many people get together and do such a good job."

Burley Irrigation District manager Evan Rasmussen designed the five-foot siphon to go under the Milner canal. Armco manager Roy Beams and his welder worked the entire night to build it.

"It's the good old USA," Roy said. "When there's a need or disaster everyone lays aside their differences and takes care of it. Everyone does their best to help when there's a need."

Another person remembered how hard the farm crews worked. Generally during the spring, farmers employed migrant laborers to help on the farms. This year they employed them to help with the canal. "Sure they were getting paid," one observer said. "But they really worked hard. They put in many long hours doing that hard, heavy work."

Men from the Pella ward, Kim Cranney's teenagers, and other groups worked on the old Cottonwood channel. It hadn't been used for years and had turned into a dumping ground for soda pop cans, iron, concrete, barbed wire—all kinds of trash. They tried to get the wire out before the heavy equipment came, Kim remembered. Cans wouldn't matter to the cats, but wire would.

"Whenever something needed to be done, somebody was there to do it," said Rex Gerratt. "I was way down at Murtaugh one time and found a pile of rocks that needed to be moved. I looked up and here come a guy with a backhoe on his way from one field to another. I told him to come and move the rocks. Pretty soon his boss came looking for him, hopping mad, because he hadn't showed up. I cooled him off, we moved the rock, and I sent him on his way."

Larry Adams was fueling equipment one night when he saw a backhoe "all balled up in some wire. I grabbed my pinchers and crawled under there to cut the wire. I bumped into somebody who was under there cutting wire, too. It was Ennis Pickett. He's not a young man, but he was there. That's the way it was. It didn't matter who they were. When people saw that something needed to be done, they did it."

"Many times I'd be on my two-way radio trying to get hold of somebody, and the person we needed would show up," said Roland Bean. "We'd say, 'How come you're here?' They'd say, 'Oh, I just had a feeling I ought to come out here.'"

Kelly Adams fueling equipment while the operators take a quick break. (Photo courtesy of Larry Adams.)

Adam's Petroleum fueled equipment along the canal. "We traveled the full length of the canal two or three times a day, twenty miles back and forth in the sagebrush," said Larry. "We destroyed one truck—beat it to death. We broke the tank, the springs, the steering, the tires. It was a good diesel truck, too.

"We went through a lot of tires. One morning I just got started and wrecked a tire on a metal fence post that had been covered up. Tires West came out as fast as they could."

Twice a day Larry's crew pumped gas for the Corps. "The Corps would line up their equipment. We'd have an hour to get them full of fuel while they were servicing them," Brett Adams said.

"It was hard work, dragging the hose, going up and down the canal banks," said Larry. "The equipment beeped whenever it was backing. The boys almost developed a phobia at the beeping, they heard it so much."

"Once late at night, the last thing I heard was a loader beeping as it backed up to me," Brett said. "I went home to try to sleep. In the middle of the night, I thought I heard beeping. I sat straight up in bed with my hand in the air all ready to fill a gas tank.

"We learned to yell and holler around equipment. We were so little, and they were so big."

Larry spent most of his time in a pickup truck looking for ways to bring fuel trucks to the equipment. He kept track of what they delivered, but did not require immediate payment. "We gave instant credit on thousands of dollars," Larry said.

Brett Adams fueling equipment; Larry Adams keeping track of the fuel being delivered. (Photo courtesy of Larry Adams.)

One night when he arrived home, the police called to tell him that his Twin Falls office had been burglarized. "The guy broke a window to get in," said Larry. "The burglar alarm went off and they caught him. They asked me if I'd like to come down and press charges. It was late at night, I had just gotten home and was dead tired. I said, 'No, just take him out and shoot him.'"

The highway department crews worked long days. A lot of people volunteered to help, remembered Oakley manager Ray Kimber, "but when we started putting in pipe, I wouldn't let anybody help. You've got to know what to do when they're swinging big pipe off a truck, where to put your fingers, and such things. Most people want to take hold of the ends. You put your fingers in the wrong place and get them cut off."

Armco distributor Roy Beams was "fantastic," Ray remembered. "I'd call him at two or three o'clock in the morning—he didn't care. He only wanted to know how he could help. We ran Twin Falls out of pipe."

The Corps told them the size of pipe to install, Ray remembered, but at Golden Valley Road, he did not follow the Corps' advice. Located half way between Oakley and Burley, Golden Valley Road was the most important east-west route.

"Bill Cranney and I didn't listen to the engineers on that one," Ray said. "They told us five-footers would work. But both Bill and I had the same feeling—they wasn't big enough. We decided to put seven-footers in instead."

They needed eight sections of pipe, each seven feet in diameter and twenty feet long. Twin Falls did not have anything that big. They could get the pipe from Salt Lake City if they could find transportation. Bill sent a truck. Winslow Whiteley sent a truck. In his notes for Friday, May 11, Terry Bingham wrote, "Semi Low Boy to Armco 333 South Redwood Road. Call Bill Cranney from Port of Entry for directions. Get back quick!"

"We had to get the pipe out of Salt Lake," Bill later said. "They were huge. Everybody thought I was crazy. Even I thought I was crazy when I saw how big they were."

"Bill had good common sense," said Kent Searle. "He knew the area, knew what water would do. He tried to be as tactful as he could be. Bill and some other person—I can't remember who—were debating on how to put in culverts. Bill said, 'We know how to do it and we're going to do it.' He knew what needed doing and would get it done."

"Whenever they had tough things to get done, Bill Cranney's the one they sicced on it," said Larry Gerratt.

Ray Kimber was determined to finish the culverts before the water came. He worked 20- to 22-hour days, his wife Sharon remembered. "I'd be asleep when he came home at night, asleep when he left in the morning. Lots of times he slept in his pickup, didn't come home at all."

The Oakley Highway District had three radios. "When they saw our trouble, Jim and Diane Schaeffer from Rupert brought more radios," said Ray. "Their radios was compatible with ours. Later, Tires West brought radios, too. I'd never met the Schaeffers before, but they knew Bill Cranney. Diane called me one day, asked if I could use a radio, and brought them over. She told the guys at their farm, 'If Kimber is on the radio, don't you try to get on.' She told me if they were on and I needed it for something to talk right over the top of them and they'd get off."

Once Ray mentioned over the radio how tired he was of eating sandwiches. "The next day here come Diane Schaeffer with a big pot of chili. She brought chili out a bunch of different times. Usually she'd bring it to Cranneys and holler over the radio where she had it."

Other people furnished food and treats. "Bea Smith brought Lynn a pie," said Paula Watterson. "She brought it here to the house. She said, I can't do anything else and I know Lynn is out there working."

Nona Jenks took cookies to volunteers from Rupert, who were working near her place. The Relief Society women from Murtaugh brought a carload of homemade cookies. People from Twin Falls brought sandwiches and hot chocolate.

Leland Baker brought "what must have been 20 pounds of fried chicken and cases of ice-cold Pepsi and fed the catskinners and everybody," Tom Tate told the *South Idaho Press*.

Lots of people helped with the lunches, remembered Lila Bowers, who was Pella First Ward Relief Society counselor. "Wonder Bread donated the bread and chips, and several grocery stores donated meat. I called sisters to make cookies for the lunches—really had good participation."

"You didn't go hungry in that valley," said Rexburg volunteer Milo Siepert. "Somebody was always there with some goodies."

"We'd take lunches in our pickups," canal director Bud Robinson remembered. "Sometimes I'd have two pickup loads of food. People wouldn't stop, kept right on working, eating on the go."

The emergency had materialized so suddenly and the response to it was so great that it was difficult for those not on the scene to realize what was happening. Oakley Stake President Rex Gerratt called Elder Hugh Pinnock, the church general authority in charge of the northwest area, to explain what they were doing. "President Gerratt kept me informed," Elder Pinnock said, "and I informed the Council of the Twelve, but I wasn't sure what was happening. I remember telling him, 'We're behind you all the way, but we don't understand it!'"

Soil Conservation officials in Boise, who did not know what the local officials were doing, feared the state office was losing control. They were not aware of the extent of the emergency, nor how rapidly the construction was proceeding. It had been only a week since they had sent someone with equipment to Oakley to measure the snowpack. They warned Kent Foster to be careful, and told him that he was not to do any work for individuals. "We are really assisting the Canal Company and county with this effort," Kent wrote in his field notes.

As the emergency intensified, the Boise officials became more concerned. Kent was not consulting them nor sending his work for their staff to check. Because he was not following proper procedures, he was leaving the SCS open for lawsuits. The state officials were particularly nervous because a few years earlier some other SCS employees had worked with the Corps of Engineers on a project and had been sued.

Kent and Burke knew they were not following proper procedures. "There is a reason for specifications and doing a good job," Kent said, "but in an emergency situation you just do it. If we had tried to survey and design everything the way we usually did, we would have been surveying while the water was running."

Burke designed a drop structure for the new canal with bays where boards could be raised and lowered to control the amount of water. By using it with a structure already in place, they could regulate the water that would go to the new canal.

"The size of the structure was above my design authority," Burke said. "I should have sent the plans to Boise for approval. There was no time for that. The big boys in Boise criticized us, but we were getting things done as best we could as quick as we could. We made the biggest effort to make sure the engineering was done the best it could be done. Survey and design-wise we had to keep ahead of equipment."

Decisions that ordinarily would have taken months were made within hours. Mike Cranney remembered seeing Tom Tate and Burke designing a structure on the hood of a pickup. Burke remembered that he and Tom were in a pickup designing the channel when "here come four to five dozers to go to work."

"The higher authority didn't have any input," Burke said. "We knew a lot of things should have been done differently, but we had no time. They told us to stop working on the canal because of the complex engineering work. Kent ignored them."

Soil Conservation Service employees J.J. McLaws, James Graham, and Kent Foster. (Photo courtesy of the Soil Conservation Service.)

Building a structure on the West Canal. (Photo courtesy of the Soil Conservation Service.)

It was so hectic in their office, J.J. McLaws remembered. "The boss was calling Boise; Boise was calling the boss. It was unreal. They was calling from Boise saying, 'You guys still doing that? We told you to quit!' and Kent was saying, 'We can't quit!' One guy even told us to 'cease and desist' from what we were doing."

"We started designing things," remembered Kent. "Someone from the state called and said, 'Don't do that.'

"I said, 'We have to do it. The guys need our help. Somebody has to do it. We can give them the practical help and that's what they need.'

"He told me to 'cease and desist' from giving them any more information."

The SCS had a line and a staff organization. "I am line, and the person calling me was staff," Kent said. "He had no authority to tell me what to do. I called my boss and said, 'If you want to tell me something, call and tell me yourself.'"

The Boise office was concerned because those in Burley were not following specifications, Kent's boss told him.

"I asked my boss if I could call the top dog," Kent remembered. "He said go ahead."

Kent talked with Stan Hobson, the state conservationist.

"He said, 'If anything goes wrong, the county could come back and sue you.'" Kent remembered.

"I said, 'If they do that, I haven't done a thing.' We didn't have anything in writing, everything was oral.

"I said, 'If there's trouble, I'll just say I didn't do anything, and they'll never be able to prove it. Besides the county won't sue. They are so grateful for everything we are doing. They signed a waiver for the Corps. If you like, I'll get them to sign one just like it for us.'

"'Do you think they'd do that?' he asked.

"'Sure they will. They're not going to sue.

"'This is a big job,' I told him. 'We're going to do it and it's going to work, and then you will want to be in on it. You should be sending me some help instead of chewing me out.'"

"The state people didn't realize how bad it was," said J.J. "They told us, 'Knock it off! Don't mess with it!' But we had to do it. We didn't have no choice. We'd have been run out of town if we'd have quit. But it wasn't that. We had to help. There was no way we could have not helped."

"I'd have resigned my job before I'd have quit," said Kent.

That night Kent went to the structure that Burke had designed and cut the SCS name off. The next day the county and canal company signed an agreement holding the SCS free from liability. Mark Jensen, the SCS area engineer in Boise, called to offer his services. Kent took him on a tour of the project, showing him what had been done, and what was still to be completed.

"Once they could see it and realized what we were doing, they said go ahead," remembered J.J. "Then they sent engineers from the state to help."

At the reservoir the water continued to rise. It flowed from Goose and Trapper creeks at an average rate of 360 cfs. Jay Gorringe released 130 to 170 cfs (6,500 8,500 miners inches) into the West Canal sending it into fields, gravel pits, any place he could find. He could not run water in the East Canal; it was full to overflowing with Birch Creek water. Reservoir water ran in a hastily constructed ditch along the highway into the Cranney gravel pits four miles north of Oakley. Water ran into some of Bud Robinson's wells. "We thought about the Environmental Protection Agency," Bud said. In a week's time the water in the reservoir had increased by 2,300 acre feet. It was eleven feet below the spillway. Most of the snow was still in the mountains waiting for warm weather to bring it down.

"We've got about 11 feet to go and it's coming up fast," Winslow Whiteley told the *South Idaho Press* on Wednesday, May 9. "If it warms up that storage space isn't going to mean much. It's possible that it could spill before [the canal is constructed on] May 15."

On Thursday, May 10, the temperature reached seventy-four degrees. The National Weather Service predicted temperatures in the eighties by the weekend.

Officials had stopped estimating how much water needed to be released. No one knew. They only knew that it would be far more than the 8,000 to 12,000 acre feet originally anticipated. It was all that low stuff coming down, the stuff that Kent had worried about in the winter, the stuff they had no records for. From his ranch on Goose Creek, Buddy Bedke was telling everyone that he had never seen so much water in his life.

The canal would not be completed in time, worried canal director Daren Critchfield. The warm weather would bring down the runoff which would fill the reservoir before sufficient reservoir water could be diverted through the canal.

Daren thought again about the height of the spillway. Raising it three feet would give them time to complete the canal, time to drain some water out of the reservoir before it overflowed. But the state would not let them do it. The more Daren thought about it, the more incensed he became. It wasn't right, he thought. The state had denied the canal company's earlier request out of hand, without looking at the dam to see whether it could support a spillway addition.

Wednesday night Daren called Kenneth Dunn, director of the State Department of Water Resources. "I told him, 'Bring your dam safety people down here,'" Daren later said. "And he could interpret that any way he wanted to.

"I told him, 'We need that three feet, or we're going to have a disaster.'"

Daren called Governor Evans. "I said, 'We are facing an imminent disaster here. We need the dam safety engineers to come down here and look at the spillway. If the government agencies are not willing to come down and help us now, when we need them, there is no use in their coming down and looking around afterwards because there won't be anybody who will talk to them.'"

Early the next morning Ken Dunn and Governor Evans flew by helicopter to Oakley. On their way they discussed raising the spillway. It was not an action to be taken lightly, Ken told the Governor. The dam had been constructed to hold a certain amount of water and take a certain amount of pressure. It was always risky to try to contain more water than a dam was built for because the pressure at the base of the dam would increase. But it would also be risky not to raise the spillway and let the water run over the top. There was danger from water running uncontrolled, and danger, too, that the water might wash out the dam from the top. The spillway had been used only once. When dams were not used, they dried out which made them more vulnerable to failure.

"We did not say much publicly about those things," Governor Evans later remembered. "We did not want to heighten the concern of people below the dam."

Besides no one really knew what the danger was. "Those old engineers knew what they were doing," said Governor Evans. "The spillway had worked once; it would probably work again. But we decided we would let water run over the spillway only as the last resort."

They met with canal board members and community leaders in "a highly emotional setting," Governor Evans remembered. The directors took them immediately to the dam and showed them what they wanted to do.

"There was such a feeling of apprehension," said Ken Dunn. "We didn't know the design criteria for the dam, didn't know if it would withstand the pressure. But we were also concerned about letting water run over the top."

Ken checked the dam carefully. He could see no signs of stress. "Some technical people on my staff were concerned and didn't think we should do it," he said. "I felt it was worth taking the risk."

"Raising the spillway was a violation of the safety of dams practices," said dam safety manager David Hollingshead, "but there was no other choice."

The design had a safety factor—only one foot of the addition would be concrete. The other two feet would be wooden planks, which could be pulled out if there appeared to be danger to the structure.

Once the state had given its approval, the canal board needed to authorize the addition.

"It was a controversial decision," said Daren. "People were saying to me, 'What guarantee do you have that the dam won't break if you add another three feet?'

"I'd say, 'What guarantee do you have that the dam will hold if it reaches the level of the spillway?'

"People told me that was the stupidest thing I had ever done, to suggest raising the spillway by three feet."

Kim Cranney remembered the controversy. "It was a scary time. There were bad feelings. Even families disagreed."

The canal board met Friday morning to make the decision. Bud Robinson described the plan. They would put one foot of concrete on the top of the spillway with channel irons imbedded in the concrete to hold another two feet of flashboards. Kenneth Dunn had said it would be a good thing to do, and would be safe.

Canal director Ennis Pickett did not agree. "I was concerned for the safety of the dam," he said. "The dam was made to hold so much water and withstand so much pressure. If you added to that, you did not know what would happen."

He did not think raising the spillway would help. He believed the canal to Murtaugh would be too small to handle the runoff. With all the wet snow in the hills, he did not think that anything could keep water from running over the spillway.

Attorney Steve Bywater told the directors that raising the spillway would increase the risk of lawsuits. They could not get the $10 million liability insurance policy which they had earlier requested. All the companies had heard about their problems; no one would issue them a policy.

"After the entire board had been informed on what was being planned and discussed, a motion was made by Daren Critchfield that the spillway be built up on the front end to delay spilling water," wrote secretary Klyda Wake. "Motion was seconded by J. Allen Woodhouse. Ennis Pickett reported that he was opposed to the plan but would vote in the affirmative to make the voting unanimous. It was so done."

Building the spillway addition was "the wrong legal decision," said Steve, "but everybody understood the risk. They felt it was worth it. They were taking legal risks that were in the best interests of the community. I was proud of the way everyone was coming together. It made me feel good to be a part of the community."

Ernie Straubhaar asked Lloyd Brothers Construction to set forms for cement and Barclay's Welding to construct the channel irons. Ernie designed a road on the water side of the spillway so Kloepfer's could haul concrete.

It was a difficult time for the canal company directors. "There were committees with more meetings than you ever dreamed of," said Winslow Whiteley. "My crew took care of the farming. I was full time on the canal. I had no sleep for three weeks—couldn't sleep when I did sleep. The telephone rang night and day."

Everybody blamed the directors for having water in the dam, remembered Daren Critchfield. "We were the most unpopular guys in town."

"The canal board was under fire with calls from quacks," said Winslow, "not quacks really, just people who were uninformed. They'd call and say, 'Is it true the dam broke?'

"'When is it going to break?'

"'I heard it washed out the spillway, heard it washed out the tunnel.'"

But not everyone was upset with them. "What it boiled down to was that they couldn't do much about it," said Oakley farmer John Martin. "They had all the gravel pits in the valley full. They did as much as could be done."

"The canal board members were very responsible people," said consultant Dal Wayment, "although at first they had a hard time envisioning there could be a problem. Through the years their biggest problem had been that the reservoir did not fill. They had a hard time believing the scenario that occurred was possible."

Dal believed that from a legal standpoint, the canal directors had no responsibility to protect the downstream residents on Goose Creek. It was not their fault that people had built on the channel. In taking leadership and trying to do something about it, they chose the harder option—one that would make them liable for lawsuits if anything went wrong.

"I was on the bridge at Golden Valley the day the water came," said Rex Gerratt. "A man from Twin Falls was standing there. It amazed him that anybody would have taken on this project. 'We'll never live long enough to see the last lawsuit over this,' he said."

Meanwhile the canal construction continued. Construction operators under the Corps' direction worked to complete the section of canal between Cottonwood Creek and Murtaugh Lake. "We want to get that ditch opened up as soon as possible," Tom told the *South Idaho Press* on Thursday, May 10. "If we don't get that Cottonwood Creek water out of the way, it could flood part of our work area and we'd be dead."

They had known from the beginning that Cottonwood Creek would bisect the canal. Tom told irrigators to try to control Cottonwood until the Murtaugh section of the canal was completed. Several men spent their full time trying to keep the water within its banks.

"It was flooding the end of our ponds, running in borrow pits, across the road," Doug Cranney said. "We installed ponds and pumps in order to use the water. We tried to run it down a hole, but there's only so many holes you can run it down."

Doug's uncle Harold Cranney had a ranch in the mountains along Big Cottonwood. He and his son K. moved cattle away from the creek as the water kept rising. Trees that had been planted more than a century earlier were uprooted and washed out. Friday or Saturday Harold called Cranney headquarters on the radio.

"The creek is as wide as the ranch," he said. "The house, shops, everything is surrounded by water. The whole ranch is covered."

Except during spring runoff, Cottonwood Creek usually had about 500 miners inches in it. Harold told them they would see 20,000 to 25,000 miners inches running into the valley by Tuesday. There were no dams on Cottonwood, nothing to stop the water. An old concrete check which had been in place for fifty years had washed out.

Near the Murtaugh end of the canal Kim Cranney was trying to straighten a caved-in culvert when Tom Tate and some other Corps personnel came by.

"Well, boys, I think we've about got her whipped," Tom said.

"I don't think so," Kim replied as he popped his head out of the culvert. "My dad and brother just called from the ranch. They said we're going to have 20,000 to 25,000 inches of water from Cottonwood by Tuesday."

The Corps people were not concerned, Kim remembered. They were sure the canal could handle Cottonwood water.

When Bill Cranney called Terry Bingham Saturday to tell him about Cottonwood, Terry did not think much about it, although he later remembered Bill's warning. The flows through the gap in the canal were becoming unmanageable, Bill told Terry. The farmers had no place to put the water. "He said Cottonwood Creek was going to boil," Terry remembered. "Those were his exact words. He said it would be like no one had ever seen it before."

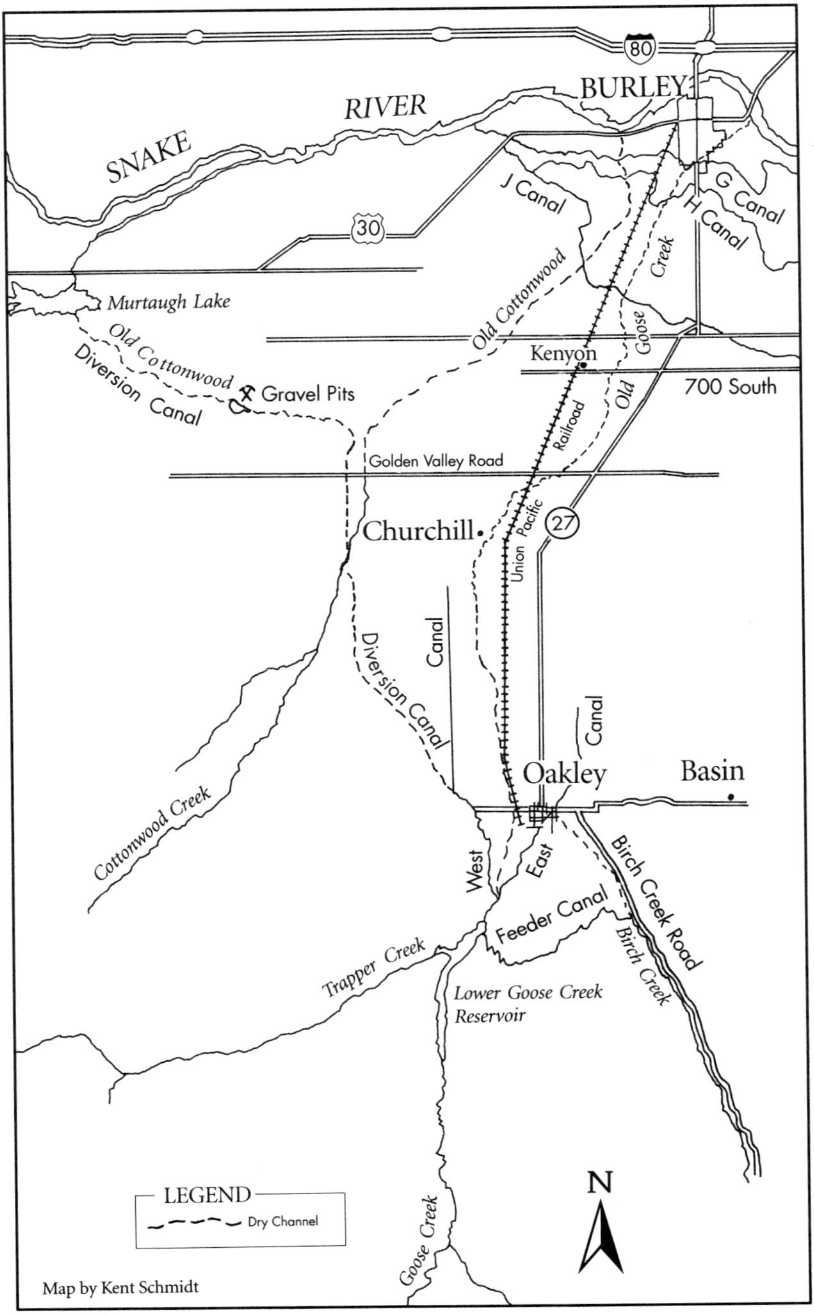
80
BURLEY
SNAKE
RIVER
J Canal
G Canal
H Canal
30
Old Cottonwood
Creek
Goose
Murtaugh Lake
Old Cottonwood
Diversion Canal
Gravel Pits
Kenyon
Old
700 South
Railroad
Golden Valley Road
Pacific
Union
27
Churchill
Diversion Canal
Canal
Canal
Oakley
Basin
Cottonwood Creek
West
East
Birch Creek Road
Feeder Canal
Birch Creek
Trapper Creek
Lower Goose Creek
Reservoir
LEGEND
Dry Channel
N
Goose Creek
Map by Kent Schmidt

Terry was not concerned. "We told him not to worry," said Terry. "We told him we would contain Cottonwood and send it to Murtaugh. We didn't really believe him when he told us how much water there would be."

The warning could not dampen Terry's feeling of accomplishment. By Sunday they would have the new canal finished from Cottonwood Creek to Murtaugh Lake; by Tuesday it would be completed to the reservoir. The weather had warmed to seventy degrees Saturday; it would be eighty-six on Sunday. But with the canal nearly finished, and work starting on the addition to the spillway, they could handle any water that the warm weather would bring down.

It had been a major effort, a magnificent effort, with so many groups and individuals all working together to complete a common goal. The canal would be finished Tuesday—exactly two weeks from the day that Tom Tate had pulled into town. Terry wondered if any other project of that magnitude had been completed so quickly, if any other group had ever planned and built a twenty-three-mile canal in two weeks' time.

County attorney Al Barrus remembered the feeling of euphoria. "With the canal to Murtaugh nearly complete, we breathed easy," he said. "We thought we'd pulled off a real coup."

Only the Cranneys and some other local farmers did not think the problem was solved. For years they had farmed with Cottonwood water. They understood Cottonwood, knew its history, and knew the problems it could cause. They understood how large the runoff from Cottonwood might be.

Lower Goose Creek Reservoir, May 14, 1984.

(Photo courtesy of Nona Jenks.)

Chapter Seven

May 13–May 15

As the water rose in the reservoir, Ernie Straubhaar supervised the raising of the spillway. The 600-foot long spillway had been built at the top of a ravine at an angle to the dam, following the ravine's natural curve. Because of the steepness of the terrain, a road for the cement trucks had to be built on the water side of the spillway.

With his front-end loader, Lawrence Burch hauled the dirt and rocks which a cat dozed out of the mountain, building the road as he went. He dumped a pile of rocks and dirt, backdragged to flatten it, then ran over the pile to dump the next load. After each load he had to back out; there was no room to turn around. Always there was danger that the loader would tip into the reservoir. A few times it started to sink at some low places. Lawrence quickly dumped his load, built up the low spot, and kept working. "It would make you wonder sometimes," he said.

Building the road took two days—longer than Ernie had anticipated. "We had to build that road two or three times because the water kept raising," he said. "We kept putting more rock on it. We thought the water would raise a foot, but then it raised two or three feet."

Meanwhile, the water in Cottonwood Creek continued to rise. On their ranch west of Oakley, Wynell and Gene Wickel watched it pond behind a road and spread toward their house. The culvert under the road could not handle the water. Although the Wickels had expected Cottonwood water to reach their place, they had never imagined there would be so much.

Leland Baker was also surprised by the amount of water. "We had never seen Cottonwood break loose like that," he said. "No one had. The old timers had never seen anything like it."

All day Saturday and all through the night Leland, his brothers, his father Wade, and other irrigators fought Cottonwood, trying to keep it in its natural channel. By Sunday morning they could handle it no longer. At nine o'clock it broke through the strip of dirt that separated Cottonwood Creek from the new canal. Whether the canal was ready or not, the water was on its way.

"I told somebody, 'Will you watch this creek for water,'" Bill Cranney said. "I turned around and, 'Gurgle, gurgle, gurgle,' here it came."

"My dad and brother were wrong when they told me we'd have 20,000 inches of water by Tuesday," said Kim Cranney. "It was Sunday when it came."

Sunday, May 13, was Mother's Day. Most people in Burley and Oakley were unaware that Cottonwood had broken into the canal.

"We attended all our [church] meetings," wrote Delma Bean in her journal. "We had a nice Mother's Day program and lessons before that. Raymond and I had small steaks, baked potatoes, tossed green salad and zucchini for dinner."

"Sunday, May 13," wrote Nona Jenks. "They have the canals about ready for the runoff. If we have ninety-degree weather and more rain, it could cause some more problems; but they feel the worst of the problem has been taken care of or is being taken care of."

At his ranch on Goose Creek above the reservoir, Morris Mitchell had spent most of Saturday trying to control flood water. He was worried about Cold Creek which flowed into Goose Creek from the east. Cold Creek had been washing down trees and other debris which had plugged a three-foot culvert.

Morris and his wife Ellen, who had been missionaries for the past two years in South Africa and Zimbabwe, were scheduled to speak in church Sunday. "13 May 1984," wrote Ellen Mitchell in her journal. "We got up early and went to check Cold Creek. It had washed completely out and even washed out some of our sprinkler system. We pulled large white pipe out of the creek. We prepared for church and when we left and went over Cold Creek, it was plugging up again."

"Cold Creek took out culvert on main road and is threatening our culvert bridge near barn," wrote Morris Mitchell. "I changed back into my work clothes and sent Ellen to church alone. I cleaned and built sides on bridge. Water finally is being carried through culvert. I changed and hurried to Sacrament meeting where we reported our mission."

Cassia County Commissioner John Adams was entertaining his mother and two sisters for Mother's Day. In the afternoon he took them to see the dam and Goose Creek. Normally Goose Creek ran through broad, flat range lands, then through "the narrows", a valley about four miles wide, and then into the reservoir. John had been there many times. Always Goose Creek had been a small stream meandering through the valley. But Sunday there was no stream. Goose Creek was a huge lake along the range lands with water running wall-to-wall through the narrows into the reservoir.

"The reservoir and Goose Creek looked like two big lakes up there with a strip of white water running in between," John said. "I didn't sleep much that night. I was overwhelmed by the threat to our valley. The extent of damage to Cassia County really hit me."

On the canal to Murtaugh that Sunday evening Ray Kimber was installing culverts at the Golden Valley crossing. This was the last crossing, the most important one, the one for which he and Bill Cranney had decided to use seven-foot pipe instead of the five-foot pipe the Corps had recommended. As soon as he finished covering the pipe, the crossings would be completed.

Ray's wife Sharon had come to Golden Valley earlier in the evening. Since the project had started, she and Ray had hardly seen each other. Carl Gooch brought her to the work site so they could visit a few minutes.

It was almost dusk when Ray saw it—the water coming from Cottonwood. "It was like a wall," Ray said, "3, 3½, 4 feet high—dirty, filthy stuff. It was coming slow, but it was coming at us. You didn't have to see it to know it was coming. You could hear it roar. When I saw it coming, I knew the culverts was history."

Lawrence Burch and Kloepfer foreman David Peterson were piling dirt around the culverts with their loaders. "Somebody said, 'Here comes the water! You better hurry!'" Lawrence remembered. "It was three-quarters of a mile away, and took about twenty minutes to get to us. When you've got a half-day's work to do, twenty minutes isn't very long."

They dumped dirt over the top of the culverts on the water side hoping to divert the water through the culverts rather than having it go on top or around them. They put rock around them to keep them from washing away. Within minutes the culverts were filled with water.

Farther down the line Kim Cranney and Slim Hewitt were inside a culvert when someone hollered, "The water is coming!" Slim started backing out of the culvert, Leland Baker remembered. "Slim is a big man, and he was moving fast. He about beat Kim Cranney to death getting out of there."

At the Buckhorn Station south of Golden Valley road where Cottonwood came out of the mountains, two backhoe operators tried to save the road crossing. "The excavators were bailing as fast as we could go," Bud Hoffbuhr said. "Cottonwood was washing down so much gravel. There was so much gravel in the pipes. The water was building higher and higher. Then just that quick the pipes were moving, and moving fast. They were floating. We were afraid they'd hit the next crossing and take that one out too, and so on down the line. It was a shame. They were perfectly good pipe. But we had to spear them, had to mash them to get them out."

For the first time, Bud understood the power of water and realized why everyone had been working so frantically to build the canals. "At the beginning, though I was concerned, I had never seen what water could do. I didn't realize it was that big a deal until I saw that water coming out of Cottonwood, until I could watch it taking a bank away. Water would erode out whatever was in its way. High speed water with volume behind it is something else."

Near the gravel pits Sunday evening a D–9 cat was in the bottom of the canal clearing away the last section of dirt. Many people remembered seeing the cat almost covered with water before it finished the job and moved out of the way.

Terry Bingham watched water flow into the first gravel pit. He had expected the gravel pits to contain the water for at least three or four days, but with the volume that was flowing from Cottonwood, he knew now that they would fill much faster. At midnight Terry went home to catch a little sleep, leaving word for the dispatcher to call him when the water filled the first gravel pit. He left, he later remembered, "thinking that I could get a good eight hours of sleep. At 1:30 A.M. dispatch called. At dawn I watched the water flowing out of the second pit. I knew we were in trouble."

Grant Wyatt, Pella farmer and former Oakley stake president, was one of the few people who was not surprised by Cottonwood. "Bill Cranney had told me Cottonwood would fill the new canal," he said. "I figured he knew what he was talking about."

Grant also knew that if Cottonwood filled the canal, another place for the reservoir water would be needed. Friday or Saturday, before Cottonwood had broken into the canal, he had gone to the SCS office and talked with Kent Foster. Together they had looked at topography maps and at the route west of Burley along the Cottonwood drain that Burke Scholer had located earlier. Grant knew about that drain—he had lived near it for forty years.

"I knew if we could get the water in the drain, it would go to the river," he said. "For maybe eighteen years, I've seen water in that drain, though not in large amounts. But about one year out of five there'd be a substantial amount of water that would run down to the river."

On Kent's map they could clearly see where the water could run along the railroad tracks, across some farmland, and then into the natural drain. "Kent and I could see that it would work good," Grant said. "The only concern was getting it past Marvin Blacker's house."

In Bountiful, Utah, canal company consultant Dal Wayment knew little about what was happening in Oakley. Dal's mother-in-law had died Saturday. Sunday he and his wife were making funeral arrangements when he received a phone call from Roland Bean.

Cottonwood Creek had filled the diversion canal to Murtaugh, Roland told Dal. He had an idea for another canal, but before he proposed it, he wanted to make sure it would work. Could Dal come and check it out for him?

Dal drove the three hours to Burley and went with Roland to the reservoir. Roland showed him the possible route—into the Goose Creek channel for a mile, along the abandoned railroad tracks, across some farm ground, then into the old Cottonwood drain to the Snake River.

"I had USGS quad maps and a hand level," Dal said. "We walked the route from the railroad tracks. A hand level isn't all that accurate, but it was good enough. We could see that it would work. We walked all the way down to the Snake River, then I turned around and came home. I had my mother-in-law's funeral pending. I may have chatted with Whiteley a minute or two, but I did not see anyone else."

After Dal left, Roland called Oakley Stake President Rex Gerratt. There was an alternate route. Dal Wayment had checked it and it would work.

Early Monday morning J.J. McLaws and other SCS surveyors arrived at the gravel pits. The day before they had surveyed the pits to determine their capacity, so the engineers could estimate the amount of water they would hold. Darkness had fallen before they were finished. They had returned to complete their work.

"I couldn't believe it when I saw all that water," J.J. said. "It would have been funny if it hadn't been so serious. There we were, ready to survey and see how much the pits would hold, and there was the water, running out already."

For the first time he understood why Kent Foster and Burke Scholer had been so concerned. "I didn't realize until I saw water running down that ditch how serious it was."

"Monday, May 14," wrote Delma Bean in her journal. "Roland, Allen, Bruce, and six or seven [of our crew] are out helping to try to get the water down to the Murtaugh Lake. So Sharla and I went to town and got thirty-five big boxes of chicken dinners to take out to feed them and others. We met Raymond out there, and he led the way from the Golden Valley Road to the end west and on north and west again into a man's yard and down his field to the old channel they are working on. Raymond took the food and waited to feed them because the water was coming close and they couldn't take time to eat. We fed Allen and came home."

At Wickels' ranch, Cottonwood water spread closer and closer to their home. A group of people arrived to sandbag. "I don't know who organized the sandbags," said Wynell. "All of a sudden they were here." Someone from the highway department cut the road. The water made a deep hole as it ran through the cut. The water flooded Wickels' corral, but it did not touch their house.

At noon Monday, water from Cottonwood reached Murtaugh Lake. Leland Baker, who had worked all night Saturday and Sunday trying to keep Cottonwood from breaking into the new canal, was now trying to keep it from breaking out. The canal was three feet deep with dirt banks on either side. No one had expected the water to be higher than the three-foot cut, but it was above it almost immediately.

"Flows from Cottonwood were estimated at 400 cfs on Sunday," said Terry Bingham. "By Monday evening it was running in excess of 600 cfs [30,000 miners inches]. As the water level rose and the velocity increased, it would carry gravel downstream. Then as the velocity decreased, it would drop its load of gravel and silt and start running sideways."

Cats waded in the water up to their radiators, pushing the dirt and gravel out of the canal. Larry Adams remembered seeing one man from Rexburg with his cat "wallering in the canal" with water up to the fans. "Water can get into the seals and damage the transmission," said Larry. "That water had all kinds of dirt and gravel in it that would eat on the seals." The man from Rexburg was "a heck of a nice guy," Larry remembered. "When there was something that the others didn't want to do, they'd holler for him, and he'd do it. He was a good operator, careful, knew exactly what to do."

"There must have been 2½ days that we had cats running in that channel twenty-four hours a day," Leland later told the *South Idaho Press*. "They were right up to their radiators in the water pushing the gravel that was washing in back up on the banks. A lot of guys had a lot of heart to run their own cats in the water like that, but just about everyone of them did it when they saw the need."

"If I needed to talk to somebody, I'd wave a cat over," Leland said. "The driver couldn't leave his machine. He'd stand on a track and we'd holler back and forth. After a minute or two the cat

Clearing silt and gravel from the Murtaugh diversion canal. (Photo courtesy of Duff Woodhouse.)

would start to sink. The water was washing the sand out from underneath—would have high centered. The driver would say, 'I've got to go.'"

The water was so powerful, remembered Leland. A boulder about four feet in diameter was in the middle of the channel. The water washed around the boulder, splashing over the top of it like white water on the Salmon River. As the water washed the dirt from underneath, it rolled a little, then stopped when it hit a high spot, then rolled again. It went about a mile. "We finally had to capture it and get it out before it plugged a culvert."

"If there was the slightest bend in the canal, the water would hit it and go to the opposite bank," said Bud Hoffbuhr. "It was going down the canal like a Z. We could watch it taking the bank away. I couldn't believe the damage water could do. I had to see it for myself before it was clear enough for me to understand."

Jim Bright and crews worked frantically to keep Cottonwood water from destroying the Milner low lift canal. The 5-foot siphon, designed to handle 250 cfs, could not begin to carry the flow. Jim installed eight 18-inch pipe over the top of the canal,

Cottonwood water crossing the Milner canal. In the center of the picture are the eight 18-inch pipe that helped to carry the flood waters. The water is pouring out of the canal spillway at the right. (Photo courtesy of the Soil Conservation Service.)

close to the water line. The water broke through one side of the canal bank. Jim opened a spillway about three hundred yards downstream to keep the water from breaking the other side.

"We had to maintain that canal," said Leland Baker. "We couldn't let it be flooded out. It was a bad place to do it. It took a lot of effort and a lot of machinery. Ralph Thornton Construction was there trying to help us get water pumped out. One of the pumps became plugged, and somebody was trying to unplug it from the side. Art Thornton jumped in the cold water and got it going again. He had no boots, just tennis shoes or whatever. 'That's the only way to unplug it,' he said when someone hesitated. 'Bale in and get the job done.' He and his crew were up to their waists in cold water trying to keep the pumps running."

"The water backed up into our sump hole," said Murtaugh farmer Melvin Tilley. "We kept jacking our pump up—ended up raising it about five feet. The water backed up into our alfalfa field, but the guys did a wonderful job under the circumstances. There was so much water—you wouldn't believe it unless you seen it. It was so swift. The road east of my place washed out. They was trying to put culverts in—the water kept washing them."

Ray Kimber worked at the Golden Valley crossing, the only crossing that had not washed away. "Ray guarded that crossing day and night," Bill Cranney said. "He was afraid it would go out. He called me and told me he thought it was going to go. I said, 'Whatever you do, just guard it. Don't get on top of it.'"

"It was such a letdown," said Ray, "so sad and discouraging to see all the culverts go after we had worked twenty-two hours a day to put them in. Within twelve hours there was only one set left."

All he could do was start over. "We had to fight all the culverts in the second time in four or five feet of water," he said. "It was awfully cold. I stepped in a hole once, went up to my neck. We got bigger culverts and wallered them in as best we could."

Roland Bean worked with Jim Bright to maintain the Milner canal. "We were dog tired," he said. "It had been a long, hard push. When Cottonwood filled the canal, it was such a letdown."

The warm weather which brought Cottonwood Creek surging down the canyon caused flooding from Raft River, Cassia Creek, and other streams in eastern Cassia County. Undersheriff Billy Crystal distributed sandbags and posted detour signs by the washed-out roads. Raft River had never run so high, washed out so many roads, nor inundated so many acres of farm and pasture land—10,000 acres in all. Ordinarily a flood of that magnitude would have commanded the county's full attention. This year it rated only a short article in the *South Idaho Press.*

"The people around Raft River thought the county let them down," said Commissioner Weldon Beck. "And I guess we did." Yet aside from sandbagging homes and cutting roads, there was not much that anyone could do. The water table at Raft River was high, the lands low and flat. There was no place for the water to go.

Above the reservoir, Mike Poulton and other ranchers were moving their cattle across Goose Creek onto their summer range. Earlier the ranchers had talked with BLM and Forest Service officials, "begging them to let us move the cattle a day or two earlier, before the flood came," said Mike. "We told them Goose Creek would cover the meadows. There would be no way we could cross when it started coming down."

The officials, who thought the ranchers were exaggerating, told them they had to wait until Monday, May 14.

Goose Creek was high on Monday. Mike herded his cattle across the Goose Creek bridge and began crossing the meadow where Goose Creek was starting to flood. "Kelly Wells, who was helping us, got too close to the water, got in a slough," said Mike. "My dad saw him go under, hollered at me. I couldn't see what Dad was looking at—both horse and rider were out of sight. The horse came up. Then the rider came up. Just as the horse was about to get out, it panicked, fell over backwards on Kelly. It did that about three times before I could get a rope to Kelly to pull him out. We had to get a tractor to get the horse out."

About four hours later, when Ray C. Bedke arrived with his cattle, the water was two feet deep across the meadow. "All you could make out of the bridge by the time Ray C. got there was the riffles," said Mike. "The water was forming a riffle as it ran over the top of the bridge. When the little calves got on the bridge, the water washed them off onto the main stream. Someone was on a bend to rope and pull them out. Bruce Bedke's horse got washed off the bridge into the creek. I had brought my two kids, thirteen and fifteen years old, to help us—Bruce had his two, the same ages. Later, I thought what a foolish thing that was to bring those kids. It was so dangerous. We all could have gotten drowned trying to save each other."

Although Mike had expected Goose Creek to have a lot of water, "it was beyond the imagination of everyone that it would be that high."

People were battling floods throughout much of the Intermountain West. One hundred fifty families were evacuated as mud slipped off the hillside and oozed into East Layton, Utah. Mud slides near Price and Toole killed a bulldozer operator and an elderly man. Volunteers sandbagged the streets in Salt Lake City as water poured from the canyons. At Salmon Falls dam southwest of Twin Falls, officials prepared for a controlled spill.

In Burley, Bureau of Reclamation officials had been monitoring the dams along the Snake River. They had not been involved with the Oakley problems, remembered Denny Davis. "We had our own flood upriver at Ririe Dam."

"Cottonwood surprised us," said hydrologist Leo Busch. "The magnitude of water was a lot greater than we as professionals anticipated. Cottonwood put up the flags for Trapper and Goose creeks. When we saw Cottonwood, that made us realize how heavy the runoff from Trapper and Goose Creek would be."

Leo knew that the next few days' weather would be critical in determining the rapidity of the runoff. More hot weather would bring the water down quickly. If the weather cooled, the snow would not melt quite so fast. Worst of all would be rain—rain on melting snow. That would bring everything down at once.

It was hot again Monday, with an early afternoon temperature in Oakley of eighty-four degrees. Later that day a storm blew in, dropping the temperature significantly.

"It has rained, hailed and the wind blew—still is," wrote Nona Jenks. "It really cooled."

That night Denny Davis slept fitfully. He was thinking about evacuating the widows in his ward and about the flood that now seemed inevitable. In the middle of the night a sound woke him. It was rain, heavy rain, rain on mountain snow. He had a hard time going back to sleep. Every time he woke up he could hear it raining.

Monday evening as the sun was going down, Terry Bingham watched Cottonwood Creek flowing into Murtaugh Lake. A light rain was falling.

"The rain matched my spirits exactly," Terry said. He knew they had failed. All the hard work of the Corps and the local people, the cooperation of the farmers and government officials, the phenomenal building of the canal in twelve-days' time—none of that made any difference. The Corps would complete its work by Tuesday, but it would not help. Cottonwood Creek was taking the canal's entire capacity. The spillway addition would give them time, but it would not keep the reservoir from running over. Burley with its schools and hospital would be flooded, the BID canals would break, farms in the center of the valley would be ruined. The water could easily spread three miles wide. The water would probably be six to ten feet deep in the old Goose Creek channel and about three feet deep elsewhere—just enough to fill basements, ruin crops, damage electrical equipment and destroy the county's tax base. They could not stop it. The only thing they could do was try to bring the water through the valley in as orderly a manner as possible.

If they could keep Cottonwood Creek from breaking out of the new canal they would be lucky. Cottonwood could easily flood all the farmers along the route and destroy the Milner irrigation system. Terry thought of their agreement with the Twin Falls Canal Company. They had promised to give them twenty-four-hours' notice and send clean water. That was a joke. Cottonwood looked like a big liquid Hershey bar.

A discouraged group met Tuesday morning at the canal company office. Canal company directors and manager Jay Gorringe, county commissioners, Tom Tate and others from the Corps, Terry Bingham, BID officials, and Oakley Stake President Rex Gerratt had all gathered to try to decide how to mitigate the disaster which everyone knew would surely be coming upon them.

They reviewed the problem that Cottonwood posed, noted secretary Klyda Wake in her minutes. "Other methods of dispersing water were discussed this morning. Some suggested turning it into old Goose Creek channel and letting it take its course cutting roads as it went. Suggested turning into old channel, cutting over to the train depot and railroad tracks in Oakley which are being abandoned and going down the tracks to the Golden Valley road and then find a route somewhere west of Burley which would avoid as much of the populated area as possible. The addition to the spillway is nearly complete and will give a few extra days. . . . All left the meeting with the idea of finding the safest way to divert the water into the Snake River west of Burley."

"It was disheartening when Cottonwood broke," Kent Foster said. "We had thought we were pretty good guys to have pulled it off before then."

But Kent also knew that building the canal to Murtaugh had not been a futile project. Even though the canal could not carry the reservoir water, it was averting a problem that few people had foreseen. The canal to Murtaugh was keeping Cottonwood water from pouring into the old Cottonwood drain and flooding all the way to the Snake River.

Kent tried to encourage the others. "I said, 'Don't be down. We did good. We saved people from being flooded with Cottonwood water.'

"I still had hope," he later said. "We could do something else. No way could we let that reservoir water go to Burley."

Terry Bingham had a myriad of things to do. His notes were terse and brief: "Channel water down Goose Creek. Direct around town. Get ASCS to survey route. Skirt to the west. Clear the track in a hasty fashion on railroad tracks. Call the Mayor. Tell press to tell every person in the way to start making preparations for flood. 15,000 acres inundated. Get Paul Massie at disaster services. We have got to direct around Burley. Sunday it will reach spillway. People on Goose Creek prepare to move. Call state on sandbags. Call Evan Rasmussen for meeting at SCS. Army bridges. Call Chuck Shadduck. Sandbag at Willow Creek. 600 S. Highway 27—fill material."

At the Oakley Canal Company office, Winslow Whiteley tried for hours to talk with one of the "higher-ups" at the headquarters of the Union Pacific Railroad in St. Louis. He requested permission to use the abandoned railroad right-of-way. The official told him they would study the situation and let him know. It would take about six months for permission to be granted.

"We didn't have six days," said Winslow.

Weldon Beck asked Steven Pearson, who owned Desert West Land Surveys and often worked for the county, if it would be possible to take the water around Burley along the 400 West road. It would be difficult to do, Steve told him. There would be a problem with sideslope and too many places where they would need cuts and fills.

Mike Cranney visited landowners in Oakley. "President Gerratt assigned me to contact people on the Oakley end to see how they would feel about diverting water along the railroad track," Mike said. "There was no problem on our end. But they stirred up a hornet's nest in the Pella area."

Cranneys, Adams, and Beans—all with large farms along the Goose Creek channel—had decided that they would not let the water come through their land uncontrolled. They talked to one of the cat owners who was working for the Corps and arranged for him to work for them when the Murtaugh project was finished.

"We told the commissioners that we would dig a ditch through our land, but we wanted the county to take control," Mike remembered.

"'You'll get sued,' the commissioners said.

"'We'll get sued anyway,' I told them. 'We filled in the Goose Creek channel.'"

Terry Bingham and Weldon Beck told reporters that water would probably spill over the dam some time Sunday. Surveyors were looking for a route to divert the water around the city. Weldon said that the route would approximate the old Goose Creek channel, then divert to the northwest somewhere near Burley. "People along the old Goose Creek drainage should be prepared to move," Terry warned.

The news stories gave the impression that the officials were more in control of the situation than they actually were. This was done deliberately.

"We were trying to play down the panic," said Burley Police Chief Leman Messley. "We had to balance between engendering panic and giving information that the people had a right to know. The local media were super. I could talk to Mike and Janet Feiler of the *South Idaho Press* and tell them, 'Here's a potential problem we can't let out yet.' They would respect that. We were torn between trying to emit a sense of security to keep panic down and the knowledge that it was one heck of an emergency."

Until Tuesday, city of Burley officials had not worried about the reservoir. Even when it became apparent that the reservoir would overflow, most city officials did not think Burley was in danger.

"I doubted it would get to Burley," said city attorney Bill Parsons. "I thought a lot would be absorbed by the ground. I didn't know how the water would get down the branches of the Christmas tree—past this canal, this little ditch, this road. It would spread out behind all of them."

Burley Police Chief Leman Messley knew better. Unlike most city officials he had grown up on a farm and had experience with water. Earlier Leman had rented an airplane so that he and Mayor Shadduck could fly over the reservoir and see the danger.

"We flew over Big Cedar Canyon," Leman said. "That was the first eye-opener I had. We were flying low. The vibration of the airplane was cutting the snow. We could see it cutting loose and slipping down the mountain. I remember thinking, 'This stuff is mush.'

"My first reaction was, 'The county's really got a problem.' Then I remembered, 'One-third of the city is built on the Goose Creek bottom. We've got a city problem too.'

"I knew we had to follow the county's lead," Leman said. "I went to Undersheriff Billy Crystal and told him, 'You give me instructions as to the responsibilities you want me to carry.' That was the last time I saw daylight."

Most Burley residents were surprised to learn that their town was in danger. Many could not grasp the urgency of the situation, even though KBAR, the local radio station, issued regular bulletins warning people to prepare. How people reacted depended mostly on their previous experience with water.

"I didn't think the water would get here," said Burley Stake President Boyd Poulton. "I had grown up in Oakley. The reservoir is such a huge holding facility. I thought, 'They can't put enough water in the reservoir to fill it. There's not enough moisture in Goose Creek drainage to fill it ever.'"

Yet in spite of his personal feelings, his stake was prepared. "We had a designated disaster coordinator, Tex Woolstenhulme. We were ready to evacuate. We knew our doctors and nurses. We had made our resource survey and tried to keep it current."

Dr. Walter Petersen, President of the Burley West Stake, at first did not think the reservoir was of concern to his stake. When it became clear that Burley was in danger, he told Priesthood leaders to warn everyone in the flood area to clear their basements and prepare to move.

Burley resident Carolyn Taylor took the warning seriously. "I knew we were near the Goose Creek channel," she said. "I called my dad and asked him if he thought the water would get here. He had been a farmer, so he knew water. He said if it went down the channel we'd be flooded. We took out flood insurance and started sandbagging."

Steve Pearson was worried about his house on Fairmont Drive which was in the old Goose Creek channel. "I met with my neighbors," he said. "Everybody wanted to know what would happen and how much water would be coming, but nobody could say how much." Steve took everything out of the lower section of his house. He and others sandbagged the homes of their neighbors who were in the most danger.

Julie Fairchild knew her house on East 17th Street was in the middle of the old channel. "I was so frightened," she said. "A guy was there trying to sell flood insurance to people in our neighborhood—really high priced. I was a single parent at the time. I didn't have the funds to take out that high-priced insurance, and was so afraid I'd lose my home."

Nona Jenks was working at her beauty shop in Burley. "During the day someone called to warn me of impending danger from waters coming over the spillway," she wrote in her journal. "I worked the rest of that day, but I was a nervous wreck. My hands trembled and I had a hard time doing a good job."

After she arrived home, she and her husband discussed the situation and decided not to move. Because their house was high, they believed that the water would go around them.

"Tuesday, May 15," wrote Delma Bean. "I went to town and got a lot of groceries etc. to help us if the flood comes our way to get prepared. Got some five-gallon water containers and a big fifty-five-gallon container for all four families with a pump. Also got lights with batteries [and] a gas grill to cook on."

"I was so scared," Delma said. "I was afraid the dam was going to break, like Rexburg. I didn't fear at first, but when I went to the Oakley dam, I could see Rexburg."

Ray Kimber, still guarding the Golden Valley crossing, had someone relay a message to his mother in Burley that she should sandbag her house, get insurance, and prepare to evacuate. His wife Sharon sent their five-year-old daughter to stay with relatives in Almo. Although they did not think their house in Oakley would be flooded, they did not want to take any chances.

"I was worried sick, knowing that the water had to go," said Bart Bowers, who farmed in the Goose Creek channel. "There was nothing I could do. I remember being real worried we'd lose the third-lift canal, lose all the BID irrigation system."

There were lots of rumors. "Somebody called me late at night," remembered Loreen Gorringe. "She said, 'Loreen are you still at the house? We heard they're evacuating and you were the first to leave.'" Loreen and Jay lived in a tall, three-story Victorian house near the center of Oakley.

"I told her we weren't going anywhere," Loreen remembered. "She said, 'Well, if you're staying, I guess I'll stay.'

"Jay said, 'You didn't tell her that we had a canoe on the third floor. When the water gets that high, we'll get in it and float away.'"

All along the old Goose Creek channel the farmers diked their houses and prepared to move. Pella farmer David Beck bought a dozer blade for his tractor. "I went to Don Reed's and built a dike around their house," he said. "We had a plan to move people out of Goose Creek. We had a telephone tree, a calling tree, organized."

But many people made no preparations. Although he lived close to the Goose Creek channel, Merlin Stock thought that their place was high enough that the water would not affect them. Nor was he concerned about the dam breaking. He had been in Montana when a dam washed out. The earthen part washed until it came to the concrete core, and then it held. He stopped worrying about the dam when he heard it had a concrete core.

Leon Adams, who had a farm at Golden Valley on the Goose Creek channel, was on a houseboat in the middle of Lake Powell in southern Utah, fishing. "We had the boat lined up to go fishing long before the flood," said Leon. "We didn't want to lose our deposit on it."

City Councilman Ken Fronk knew that Goose Creek channel ran about a hundred yards east of his home, but he made no preparations. Ross and Karen Taylor, his neighbors, moved their food storage and carpets from the basement. Bob Bray sandbagged and diked around his house and bought a sump pump. Ken did not have a basement, and did not think he would get wet.

Pella farmer Fred Hawker talked to City Councilman Dale Doman. He said to Dale, "You've got some water coming your way."

"It'll be no problem," Dale replied. "We've got it all figured out. We'll take it down the channel and west of Burley."

"The city council talked a lot about what we'd do if the water came," Ken said. "We figured at the first-lift canal we'd have to move it around the city, but we didn't know how we'd handle it. We made no concrete plans."

The Salvation Army brought volunteer workers from Twin Falls and set up a portable kitchen at the Burley service center. Welfare coordinator Wayne Heinemann told *The Times News* that they had medical volunteers and Search and Rescue volunteers ready to come to Burley.

City employees filled ten thousand sandbags and ordered thirty thousand more. They filled bags each night "until it was so dark we couldn't see," said street supervisor Olene Spurgeon.

"We sent out word over the radio that we needed volunteers. I was amazed at the number of people who came from Minidoka County—of course people from Cassia came too, but those in Minidoka were not in any danger. We had older people, kids, people of all age groups out there. A lot of ladies came bringing their little kids three and four years old. I moved sand in a different place for the kids, where they'd be out of the way of the trucks. They had their own little sandpile, kind of like kindergarten."

Some people came day after day, Olene remembered. A few brought their lawn chairs. "They'd fill sandbags awhile, sit down and rest a little, then go back and fill again.

"A vagrant came by—not a vagrant really—a homeless gentleman. He filled sandbags, worked hard for three weeks. For three weeks he lived there at the service center. At the end he said, 'It looks like I'm not needed any more,' and left. He was probably in his mid-fifties. He had a motorcycle, was a wanderer."

Greenline Sales brought a big truck for hauling sandbags; Volco Builders, a smaller truck. Olene spent most of his time supervising city trucks and filling sandbags.

"My wife said to me, 'When are you going to sleep?'" he remembered.

"I said, 'When it's done. I'd work as high as about thirty hours at a time, then sleep, then get up and go again. My wife would come home from Ore-Ida where she was working. She'd say, 'Well, I see he hasn't been here yet.' She'd get the kids and come out. We'd talk and fill sandbags."

Undersheriff Billy Crystal updated county evacuation plans. The town of Oakley was high enough that even if the dam broke, Oakley residents would be safe. People near Oakley could go to Oakley churches and schools, those south of Burley to View and

Unity churches. Some people could cross the bridge to Minidoka county. If the dam broke, law enforcement officials would alert the public with announcements on the radio and drive through town with loudspeakers. Finally they would go door-to-door to make sure that everyone heard the instructions and knew what to do.

Police officer Kirby Harkness planned for the evacuation of Burley. "We had to get people across the bridges in an organized way," said Police Chief Leman Messley. "Kirby came up with a plan for people on certain streets to go certain ways. We put it in the paper. It was a worst case scenario, but we had to plan it that way."

Since the beginning of the crisis, Buddy Bedke, from his ranch on Goose Creek, had relayed information about the depth of the snowpack and the extent of the runoff.

"When the whole thing started to come down, Buddy Bedke was about the most reliable source that I had," said Leland Baker. "He had land up there in the high country. He told us he'd never seen so much up there, that it would be here in X amount of days."

Tuesday was school board election day. Reine Robinson worked with the election at Oakley High School. "It was raining," she said. "The people were so frightened. They said the dam would wash. It wouldn't be able to hold when all that water hit. People from the Goose Creek area were coming down, scared to death. Buddy Bedke came in and said, 'I'm getting out of here!' He checked his kids out of school and left."

"Buddy Bedke was saying, 'You've got more than a creek. You've got a tidal wave here,'" Richard Hardy remembered.

"I thought the dam would break," said Buddy. "I took all my family to the ranch above the dam, where they would be safe."

For Oakley students, Tuesday was the last day of school. Officials cancelled the remaining six days of the school year and asked older students to report to the high school to fill sandbags and help people along Goose Creek channel who needed assistance. Twenty families along the channel were told to evacuate.

Mike Campbell, hydrologist technician with the U.S. Geological Service, arrived to measure the flow from Goose Creek. Especially during high water, the USGS liked to hand-measure flows so they could make adjustments on their gauge readings. "We were running all over the country," Mike said. "There were only two of us in the Twin Falls office, and water was everywhere. Besides Oakley, there was major flooding at Raft River and Salmon Falls."

The gauges on Goose, Trapper, and Salmon Falls creeks had tape on which the flow heights were recorded. When Mike Campbell read the Goose Creek gauge Tuesday, he found that water from Goose Creek was flowing at the rate of 1,320 cfs, almost double Monday's gauge reading of 770 cfs. An additional 150 cfs was flowing from Trapper Creek, plus an unknown amount from the Birch Creek feeder. About 3,000 acre feet had entered the reservoir in twenty-four hours.

Along the reservoir spillway, Max and Gary Lloyd set forms for the spillway extension. They drilled into the rock on the spillway, then pounded in half-inch steel rebar to anchor the forms.

"Lots of people would drive up and look," Max said. "I bet there was fifty cars a day coming up and observing. Townspeople would come up to us and ask what we was doing. We told them, 'We're doing what we've been asked to do by the canal board.'"

"We got a lot of comment from people," said Denver Altom. "Lots of people were saying it would never work. Some people thought we were crazy."

People could see water seeping through the spillway, coming out the other side. "When they seen the water squirting out of that old concrete, they was really concerned," said Max. "I had my doubts whether it would hold or not when the water started squirting."

About ten volunteers helped set forms. Max's two high school boys worked on the spillway after they were dismissed from school.

"It was like an adventure at first," said high school junior Delbert Lloyd. "But when we saw how fast the water was rising, it became real. We were watching a juniper in the side of the mountain. It got covered up pretty quick. Then we stuck a stick in the mud and watched that. We could actually see the water rising up that stick."

"It was a spooky feeling," said Max, "watching the water raise, wondering if we'd get it done. We'd put rocks and stuff and sticks on the banks to see how much the water was raising. It raised about two inches in four hours."

Kloepfer trucks began pouring concrete Tuesday afternoon. In addition to driving on the water side of the reservoir with the

West end of the spillway looking east, May 14. The steel posts are in position on top of the old concrete spillway. New concrete one foot high will be poured, which will anchor the posts. (Photo courtesy of Nona Jenks.)

water constantly rising, the drivers had to come down a hogback with top-heavy concrete mixers, heading directly into the reservoir.

"Some drivers who had been driving all their lives were pretty scared," Rex Gerratt said.

"It was beyond the normal call of duty for the men, but they did it," said Ernie Straubhaar. "Except for one man. He refused to go down there. He said he got claustrophobia at night and he couldn't do it."

"There was a lot of stress and pressure," said Max. "We had twenty-four hours to get the trucks in and out. We worked all night, had to have lights. The guys was tired, the truck drivers was tired, it was raining and slick. It rained half the night on us. I was nervous about the trucks going across the spillway. There was eleven to twelve yards of concrete on one truck, a lot of weight, about forty-two thousand pounds. If it was too close to the edge it could have gone into the reservoir. It don't take much to tip one over if they hit a soft spot."

View of the spillway from the dam looking southwest, May 14. Some of the steel posts are in position, others ready to be placed. (Photo courtesy of Nona Jenks.)

"We were working across the way and saw a truck coming down," said Randy Hardy. "Max said, 'Boy, I sure hope his brakes don't go out.' Then his lights went out—flickered a little and went out. He was going downhill, it was steep and curvy. He couldn't stop, his brakes wouldn't have held. He opened the door. There was enough moonlight that he could see below him the road he was driving on. If he hadn't been such a good driver, if he hadn't been driving for so many years, he would have been in the reservoir."

"He got out of the truck and he was white," Max remembered. "He said, 'I'm not going to come back here again.'"

Twenty-six hours from the time they started, the concrete was poured. "We finished just in time," said Ernie. "The water was at the bottom of the tire of the last truck as it poured the last of the concrete."

"They couldn't have got another truck in there when the last truck went out," said Max. "Another truck would have tipped over into the reservoir."

Terry Bingham had heard rumors that if they raised the spillway, someone would take out the planks. He did not think the threat was serious—"the talk came from a few of the guys who stayed too late at the Oakley tavern"—but he could not take chances. The sheriff's office posted guards at the reservoir and at the gates.

Doug Adams, volunteer reserve deputy with the sheriff's department, was one of the reservoir guards. "It was amazing to sit there and watch that water raise up," he said. "They had marks on the spillway so you could actually see it raise. We had orders to shoot if anyone came up there and started fooling with the dam—that part was scary. Nothing happened. The threat to the dam was all just talk. I figured it was—I knew some of the guys doing the talking—but we couldn't take any chances."

Reserve Deputy Kent Searle also guarded the dam. "Before the spillway extension went up, we pretty much watched," he said. "We didn't try to chase people off at first, just watched to make sure no one did anything strange. We blocked the road after they built the extension. I'd make a mark when I'd go up there. During my shift I could see that the water had come up. I knew that water was going to go over the dam. I could see it, all backed up there."

But Oakley Stake President Rex Gerratt was determined that the water would not go over. "We will do everything within our power to see that not one home is flooded," he had said at the beginning of the crisis. He had no intention of giving up now.

He called a meeting for bishops and ward and stake emergency leaders.

"President Gerratt reviewed the progress to date," wrote stake clerk Donald Clark in his minutes. "The Oakley Canal Co. is engineering a new canal to the Snake River so a managed diversion can be handled instead of turning the excess water loose. The Stake Presidency firmly recommends we demand a managed turn out. It may be necessary for you Bishops to obtain the necessary easements. It will be necessary for all Bishops to be on call at any hour during the next ten days because changes are taking place each minute. . . . Put all Quorums on alert, and 100% cooperation and support is expected."

They discussed specifics—sandbags to be filled, residents to be evacuated, help to be given to widows and the disabled. "President Gerratt will be on call at all times. Ron Hepworth will supervise days and Wayne Mullen will supervise nights. . . . We can expect 40,000 inches of water to be released by Saturday or Sunday and could run for 30 days."

That same evening the commissioners, canal board, and SCS officials met at the canal company office. They talked about where the water would go, how they could divert the water around Burley, and what plans they should make.

"I was assigned to look for alternate routes and give my findings to Terry by noon," Kent Foster wrote in his notes. "Called Grant Wyatt and told him to be there at 7 A.M."

Terry Bingham did not attend the church or canal board meeting. He was advising residents along Goose Creek channel to evacuate. The water would flow down Goose Creek, he told them. They would release the water into the channel behind a V-shaped ditcher. He scheduled a meeting for the next morning to prepare.

About 10 o'clock that night Pella farmer Marvin Blacker saw County Commissioner Weldon Beck and "one of the SCS boys" on the 400 West road by his farm. He was surprised to see them, since his farm was not on Goose Creek channel.

"They said they were going to put a canal down my road," Marvin remembered. "I said, 'It won't fly, boys.'

"They wanted to know why not.

"I said, 'Because you can't dig a canal through solid rock, and that's what's under this road.'

"They were grasping at straws," Marvin later said. "They were searching for a place where the least amount of damage would transpire. It's a lot cheaper to rebuild a road than farmland."

Sandbags on pallets at Joe Beck's farm.

(Photo courtesy of Gwen Beck.)

Chapter Eight

Wednesday, May 16

Wednesday at 6:00 or 7:00 A.M. thirty-five to forty landowners were waiting outside the SCS office in Burley. They crowded into the conference room to hear Terry Bingham's report.

There would be a flood, Terry told them. Water would reach the top of the spillway extension on Sunday. The commissioners had decided to let the water flow down the old Goose Creek channel, releasing a small amount of water at first. Soon they would need to match the outflow with the inflow.

Marvin Blacker later remembered how sorry he felt for all the people along Goose Creek channel whose farms and homes would be flooded. He told the group he would be glad to help in any way he could.

Meanwhile at the law enforcement building, city and county officials met with Tom Tate. "People were incredulous," said Undersheriff Billy Crystal. "They were saying, 'This can't be happening. It can't be that bad.'"

"Tom Tate did some drawing on the board," Police Chief Leman Messley said. "I was still in the denial process. I didn't quite believe him. It was a roomful of disbelievers. Tom was telling us the water was coming here, and yet he was qualifying his statements, too. He was saying the potential is here. We believed him, but were still thinking, 'Goose Creek has never run over. The dam has never been full.'"

Someone suggested sandbagging along Overland and Parke avenues. Officials in Salt Lake City had controlled flooding by sandbagging their streets. Many people had seen it on television.

It seemed the only feasible plan, yet even as they discussed it, Tom knew it would not work. The street in Salt Lake had a good slope along a natural drainage. Overland and Parke were flat, with no relationship to the natural drain.

"They talked about letting the water run," said John Adams, "but I remembered seeing that torrential water above the reservoir. Most people didn't see that."

"They were going to let the water run," said Grant Wyatt. "I was dead set against that. It would have spread out a mile or more in places. They had no plan to get it through Burley except by sandbagging the streets. They could not run it down Overland. Those who thought they could did not realize how much water 40,000 to 50,000 inches was. There was no way they could control it.

"I told them we ought to look at what we could do. Perhaps there was a way we could avoid bringing it into the city. We could look at it, and see if there were other alternatives."

Terry Bingham, Weldon Beck, Al Barrus, and others drove to the place where the Goose Creek channel bisected the third-lift canal. Grant Wyatt and Rex Gerratt decided to drive along the route. Some people remained at the SCS office looking at maps.

Bill Cranney heard about the plan and called his brother Mike. "Terry said they are going to let that water go," Bill told him. "Get down there and tell the commissioners, 'Don't let it go!'"

Roland Bean heard about the SCS meeting and drove to town. "I found Grant Wyatt and President Gerratt at the shopping center parking lot," he said. "They were going to take a tour."

They drove the route in a car caravan, picking up farmers as they went. One person thought there were about twelve cars altogether. Another thought there were fifty. They met the county group at the third-lift canal.

"[We] stopped in front of Ira Coltrin's residence at 200 West 460 South," Terry Bingham told the *South Idaho Press*. "Grant Wyatt and a delegation from the Pella area arrived with a plan."

The route that Grant proposed—along the railroad tracks, across some farmland, and into the old Cottonwood drain west of

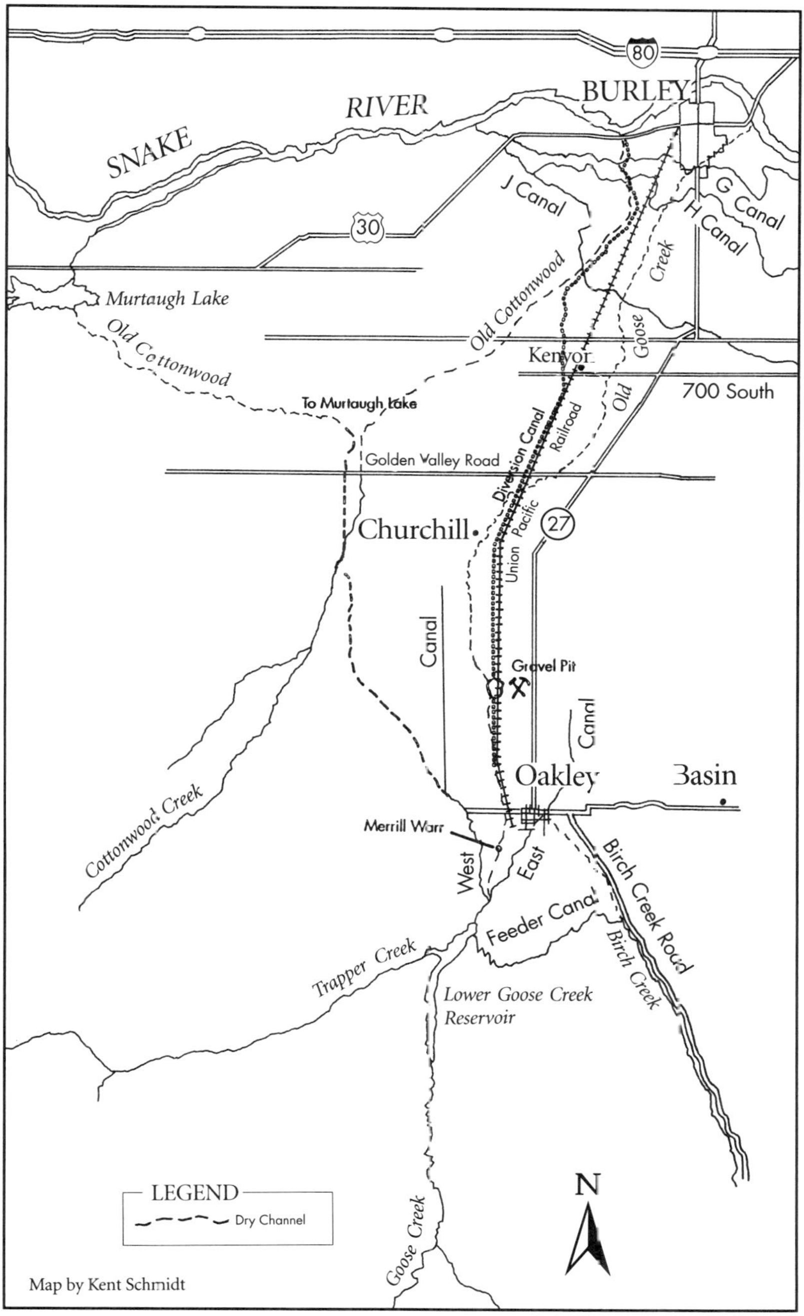
80
BURLEY
RIVER
SNAKE
J Canal
G Canal
H Canal
30
Creek
Murtaugh Lake
Old Cottonwood
Old Cottonwood
Goose
Kenyon
700 South
To Murtaugh Lake
Old
Railroad
Diversion Canal
Golden Valley Road
Pacific
Union
27
Churchill
Canal
Gravel Pit
Canal
Oakley
Basin
Cottonwood Creek
Merrill Warr
West
East
Birch Creek Road
Feeder Canal
Birch Creek
Trapper Creek
Lower Goose Creek
Reservoir
LEGEND
Dry Channel
N
Goose Creek
Map by Kent Schmidt

Burley— was the same one that Burke Scholer had earlier found on his topography map and that Dal Wayment had surveyed.

"Grant Wyatt at the canal bank first proposed that route in my presence," Terry later said. "They were excited. They thought it would work."

"At first I didn't think it could be done," Weldon Beck said. "I was skeptical. I thought it was uphill from the grain elevators to the drain."

In his house near the third-lift canal Rick Hirst could hear them arguing.

"It turned into a real shouting match," said Roland Bean. "The commissioners started to tell us they were going to open up the channel and let the water come. We said, 'No way can you afford to turn that water loose.'"

"We wanted to dig," remembered Mike Cranney. "The county said, 'Forget it. We'll go the natural channel and cut it slow.'"

It was stupid, someone said, to try to run water through town past schools and the hospital. They could never do it.

"We can't let that water get into town," said Grant Wyatt. "We'll never control it in town."

The county was concerned about liability, Mike remembered. County attorney Al Barrus was there, telling them of legal problems.

"President Gerratt said to the commissioners, 'Dig the ditch,'" Mike remembered. "'If you don't dig it, we will.'

"The commissioners said, 'You'll get sued!'

"'No, we won't,' said President Gerratt. 'Nobody will get sued. We'll talk to the farmers. We'll get them to agree.'"

Then they began arguing over whether it could be done in time. Somebody said, "You guys will get that halfway dug, and we'll still have to let the water go."

"We cornered Ernie Straubhaar and said, 'Ernie, is it possible?'" Roland later remembered.

"Ernie said we could get her done," remembered Mike. "It was a matter of machine hours. If one cat can dig a ditch in a thousand days, a thousand cats could dig a ditch in one day."

"I knew we could accomplish it," said Ernie. "X number of machines could do it, and we could get those. I was concerned about how the ditch would react, whether it would handle the

Oakley Stake President Rex Gerratt. (Photo by J Malan Heslop, Church News.*)*

water. My biggest concern was, were they accurate on their measurement on the amount of water we would get?"

Grant Wyatt and President Gerratt were influential, remembered Roland, but it was County Commissioner Weldon Beck who had to make the decision.

"He said, 'If Ernie thinks it will go, we'll do it. I don't give a damn about the liability, we've got to do what's right. We're not going to jeopardize this farm ground and the city. We won't let it go without a fight.'

"I have deep respect for Weldon Beck," said Roland. "He was gutsy enough to do it."

They scheduled another meeting at the law enforcement building at 10:00 A.M. Everyone scattered to alert the fifty-three landowners along the route.

Weldon Beck was one of the fifty-three landowners. The proposed route was near his house on land that he and his sons farmed. The water would go through Grant Wyatt's farm, "though only through the bottom end of one field." It would be within a few hundred feet of county attorney Al Barrus's home.

"From the county point of view, I was relieved when they made the decision," Al said. "I knew there would be the potential for farm ground liability, but it would save schools, the hospital, homes. It was better to cut through some farm ground than take out the town.

"At the same time, it caused me a lot of personal consternation. I knew the water would be coming next to my house. I didn't say anything, but I called my wife and told her the water's coming. Then I went down and took out flood insurance.

"When disaster strikes, you feel strong," he later reflected, "but when the water is headed to your house, it's a different matter. When it touches your own home and family, you don't feel so brave."

Canal directors Winslow Whiteley and Ennis Pickett and secretary Klyda Wake had property that the canal would cut; it would run through Cranneys' and Beans' properties and the farms of others who had been working hard on the Murtaugh canal. They were not concerned. All except Klyda had land on Goose Creek, so would have been flooded anyway.

"The benefits we would receive far exceeded the damage it would do to us," said Ennis Pickett.

But many Pella and Star landowners had not expected the flood to affect them. A little ridge of higher ground separated the old Cottonwood drain from Goose Creek. Those who lived west of the ridge did not think they had anything to worry about.

Jack and Cleo Gochnour (Gahk′ ner) lived near the Star Ward church, where they operated a 160-acre dairy farm with two of their sons. Jack's father had started the dairy in 1912. The Cottonwood drain came through "the full width of my 160," said Jack. "It started in the south corner and came through the middle. It was small, like an irrigation ditch. Some years we'd have water in it from the snow melt." That Wednesday morning he learned that the proposed new channel would come along the drain.

Jack's son Lane operated a dairy farm near the place where the Cottonwood drain ran under Highway 30 and the railroad tracks. Next to Lane's dairy were the Harris' and Hanks' properties.

Originally Gus Hanks had cleared the land from sagebrush and farmed it. In 1984 his son David Hanks lived in the old home and farmed part of the land. His daughter Marian Harris and three of her children had homes and a small farm on the other part, though Marian's husband Ed also worked full time at the

Ore-Ida processing plant. Gus Hanks, still in good health in spite of his ninety-one years, lived with Marian and Ed.

Having heard nothing about the proposed new water route, Marian was surprised when she was asked to come to a meeting. "Of course we knew about the reservoir problems," she said, "but we didn't think it would affect us. There's a natural berm between us and Goose Creek. They told me to come to the court house. Ed was at work. I didn't know what to think when I was asked to come."

In the Pella area south of Star was the Randall dairy farm. Grant Randall's father had homesteaded there in 1909; later Grant cleared additional land from sagebrush. Grant farmed first with his father; later he operated a dairy with his sons Richard and Steven. Steven's wife Julie was the daughter of Jack Gochnour. By 1984 Richard and Steven were managing the farm and dairy. Grant Randall, seventy-four years old and in poor health, was not able to do much.

The Cottonwood drain ran between their homes and the dairy barn. "It was a real beauty spot," Julie Randall said. It had an irrigation pond surrounded by cottonwood trees. The children went boating on the pond; ducks built their nests along the bank.

The Randalls had heard rumors that reservoir water might be coming their way, but they had not paid much attention to them. Wednesday morning Weldon Beck asked them to come to a meeting. "This time we believed him," said Julie.

Fred and Joan Hawker's farm was on the Cottonwood drain south of Randalls. The Hawkers had lived at Pella since 1953. Fred had worked for another farmer for five years, then had purchased a small farm on the Goose Creek channel. In 1981 he and Joan rented out their old place and moved to their new home and farm.

Fred, who had attended the meeting at the SCS office, was feeding cattle at his Albion ranch thinking that the water would cross his rented property. Joan was home when the phone rang. "The water's coming your way," someone told her. "They're having a meeting at the law enforcement center. Get Fred to it."

All along the route, landowners heard the news. Some had large, thousand-acre operations, others only a few acres. Martha and Dorothy Lattimer, sisters-in-law, both widows, had forty acres each. For some the canal would come through the side or end of a field, doing minimal damage. Others would lose several acres of crops and have their whole farming operation disrupted. For some it came as a surprise. Others had been working for weeks to avert flooding. Marvin Blacker, who left the earlier meeting feeling sorry for people on Goose Creek, now learned that the canal would cut through the middle of his farm.

Val Robins was helping his neighbors on Goose Creek move things out of their basements. He had helped with the construction of the Murtaugh canal, having gone with the Pella group to secure culverts and clear trash from the drainage area. "I was still happy and joyful at that time and willing to help," he said.

He was helping to lift a piano from a basement when Willes Blauer arrived. "He said they'd changed their minds. The canal was going to come through our farms," Val later remembered. "At that point we forgot about our neighbors, and became concerned about us. We had been sorry for the people along Goose Creek, and then when it was coming our way, we got upset at them."

At 10:00 A.M. property owners, canal directors, government officials, church leaders from all denominations—seventy to a hundred people in all—crowded into the conference room of the law enforcement building.

"The room was jammed full," remembered Police Chief Leman Messley. "We got involved in bringing extra chairs in—all the chairs you could put in there."

Some people stood at the back of the room. Others stood outside in the hall.

Terry Bingham conducted the meeting. Kent Foster traced the route on a map. Hydrologists, people from water resources, surveyors—someone from each government department—gave information.

Weldon Beck addressed the group. "We tried to let them know there would be more damage than they ever realized," Weldon said. "We had to buckle down and ask for cooperation. We knew we were responsible for the welfare of the citizens."

"There were two major problems," said Burley West Stake President Walter Petersen. "First was to convince the people in the city that they were in danger—I didn't realize the problem initially. The second was to determine what to do to minimize damage and property loss.

"There were four groups involved. There were the people along the route who would not otherwise have been hurt—some were afraid their homes would be washed away. There were the county officials. They were interested because of their civic responsibility. Third were the Burley people, and fourth, those people who farmed in the creek bed. Some of the groups were overlapping."

Evan Rasmussen, Burley Irrigation District manager, knew that water in Goose Creek would destroy the BID canals, leaving twenty thousand acres west of Goose Creek without irrigation water. Yet, when he heard the plans for digging the new canal, he was skeptical. "I thought it was an ambitious project," he said. "I wasn't sure we could pull it off."

As soon as he heard the proposal, Evan called Armco distributor Roy Beams in Twin Falls and asked him to come to the meeting. They would need culverts for places where the channel bisected the BID canals. The meeting was in progress when Roy arrived. Roy and Evan sat in the back designing culverts while the discussion swirled around them.

Afterwards, people remembered different things about the meeting.

"Everybody wanted a canal," said Weldon Beck, "but nobody wanted it on their property. Still, there was very little hostility. They realized that the canal was the lesser of two evils."

"There was lots of discussion, lots of concern," said hydrologist Leo Busch. "It was well coordinated and very informative. Certain people felt it was an overreaction. They were questioning the need of it."

"Some people were saying, 'Just let the water go. It won't bother anybody,'" Don Clark remembered.

"Grant Wyatt made the suggestion for the route," said Roland Bean. "Grant soothed feelings. He was the one the guys respected, the one who could influence them."

"It was pretty much decided by the commissioners that the canal needed to be dug," said Lane Gochnour. "I don't remember them having much trouble talking us into it. We all realized if they didn't build it, Burley would be in big trouble. It was a real numbing, shocking deal."

"There was very little opposition," said canal company director Ennis Pickett, "a little opposition, but not much." Although he was relieved that they were going to do something that might save the city of Burley, he was not optimistic. "I thought it was impossible to build that large a canal in 3½ days," he said, "an absolute impossibility."

"All the neighbors were there," said Jack Gochnour. "There was not much opposition. When we found it would flood schools and the hospital, there was not much opposition."

"They said, 'You've been chosen to volunteer. You're it,'" remembered his wife, Cleo. "There was no big quarrel. Everybody said, 'If we have to, we have to.'"

"They had good speakers," said Jack. "The men who spoke were all well respected."

Marian Harris remembered "how concerned and stirred everyone was," but she did not remember who spoke, what was said, or who was there. At the meeting she heard for the first time that her place would be affected.

"It took awhile to register with me," she said. "It took awhile to realize what would happen to us. I was kind of in shock. I remember that Mrs. Leonard King was there. I don't know why I remember her—maybe because she was one of the few women among all those men."

"They pulled maps out," said David Beck. "It was an eerie feeling. We were willing to do what we had to do."

"They were telling the plan," remembered Police Lieutenant Jim Higens. "Most farmers were saying, 'If we've got to do it, we've got to do it.' It truly amazed me—they were so good about it. I have a lot of respect for those farmers."

"One or two people were quite upset," remembered Mike Cranney. "They were saying, 'Why go here instead of elsewhere?' A lot admitted it was better than hitting schools. Only one person was really upset, but he was willing to do what was best."

"Mike Cranney talked about divine intervention," said President Petersen. "He felt that the decision to build the channel was inspired."

"What hurt was to see the faces of those guys whose land would be cut through," said John Adams.

"The farmers would lose crops, have their irrigation systems disrupted," said Bill Crystal. "I was impressed with how willing most were to deal with it in such a calm manner."

"There were good things and bad about the meeting," said Grant Wyatt. "There was lots of concern. Who's responsible for lawsuits? Attorneys saying we can't do this or that. Everything you do makes you liable.

"Wade Baker stood up and said, 'We can't worry about what we can't do. We've got to worry about what we can do. We've got to do something. It's better to do something than sit around and wait until the water hits us.'

"That broke the log jam," said Grant.

"Everybody was so tense and worried," remembered county attorney Al Barrus. "People were being asked to sacrifice their farms and future to save people they didn't know. The prosecutor's office was asked about the legal consequences. You'd like to take six months to make a decision like that. We made it in five minutes."

"There were some emotional expressions and differences of opinion," said President Petersen. "Some expressed an emotion without knowledge. There was some feeling that people were going to lose their whole farms. There was some militancy, some people saying, 'We're going to do it no matter what.'"

"There was so much emotion," Don Clark remembered. "People were saying, 'You're protecting the city, but if it's coming through my property, who's going to protect me?'"

Terry Bingham remembered someone saying, "You're going to save the big guys in the middle and us little ones on the side will get squished."

"They asked us how we felt about it," said Willes Blauer. "We told them right out. We didn't want it on our property."

"It was not a harmonious meeting," Fred Hawker said. "People were trying to cast the blame. 'It was the fault of the Oakley reservoir people because they didn't have an outlet.' But they did have an outlet—it was Goose Creek. They had a spillway.

"'It was the fault of the Oakley farmers for not watering in March and April.' But it was cold and freezing and water would have ruined their fields. I wouldn't have wanted to water either.

"I seen then there was no use hiring a lawyer," he said. "There was nobody to blame."

"There was a bunch of scrambling and fighting," said Marvin Blacker. "The city, the county, Burley irrigation, the highway were all fighting and looking out for themselves. The city council was thinking an inch when they needed to be thinking a mile. They didn't understand how much water there was. I was upset and depressed. We didn't know how long the canal would be there. We thought initially it would stay there forever."

The canal would cut through the middle of Marvin's farm. His land was flat, with only one inch difference in elevation from the top to the bottom of his field. His house was low. If anything went wrong, he would lose everything.

The Cottonwood drain ran between Randalls' house and dairy barn. "The real shock was that it was coming in three or four days," Richard said. "Nothing was arranged. We had livestock on both sides. There was no access for the milk truck. We had to move all the feed, bales of hay, corn silage. There was so much uncertainty about how much water would be coming. It was hardest on Dad—not knowing whether he could ever build it back up again."

Attorney Steve Bywater was concerned about the legal consequences. "It was a scary meeting from a lawyer's point of view," he said. "It was confused. It was unclear who, if anyone, had any authority. Votes were taken, decisions made as a group, but nobody had authority to make the decisions or take votes. It was like Athens—a pure democracy.

"The religious leaders were the most adamant. They were going to do something. It was a curious blend of democracy and theocracy. I would listen to what was said, and know that later we would have to go through with the commissioners, who had the authority, and make decisions as to what we could and couldn't do.

"People would point to us and say, 'We can do this, can't we?'

"They wanted us to say yes, but all we could say was, 'We don't know.'

"It's unnerving to a lawyer to have to make wide-ranging decisions involving all kinds of potential legal problems without having a couple of months for research," said Steve. "Shooting from the hip on issues of that magnitude scares you to death."

Terry Bingham knew that if there were liability problems, the Corps of Engineers would not help them. "Tom Tate said he would do his best, but he thought it was doubtful the Corps would participate," remembered Terry. "He advised us to go ahead anyway."

As a former minister, with training as a counselor, Leman Messley was aware of feelings—of fear, anger, suspicion. "There was a definite point in the meeting when someone stood up and began to bring things into focus," he said, "even among those who disagreed as to where to run the water, the ones saying, 'Not through my farm.' There began to be a more human concern than a territorial concern. Up to that point people had been saying, 'What do I care if city people get washed out,' and the city was saying, 'Do what you want out there, just don't bother me.' One or two inputs began to turn the corner for everybody. Anyone with political power wouldn't have been able to do it. The sheriff couldn't have done it. As police chief, I couldn't have done it."

Terry Bingham later described the meeting in the *South Idaho Press*. "Questions were asked," Terry wrote. "Who would pay for construction; who would stand the damage to the property that would otherwise be unaffected? The meeting ground on. No one had the answers."

Until that point Oakley Stake President Rex Gerratt had not said much.

"[He] stood and the room fell silent," wrote Terry.

"'Brethren,'" he said, "'we are committed, we have been from the start, we must not let any home get flooded. We've come too far to quit now. Time is wasting, we must get started. These landowners. . . will be taken care of. I promise you and them, we will not let them down. It is true we don't have all the answers now, or how it will all work, not the legal ramification, or where the money will come from. We'll have to solve some of these problems later. But, this plan will be millions of dollars cheaper than allowing the water to spill and run through our county, ruining our farms and homes and the city of Burley and destroying our tax base. Now let's give Ernie the go ahead. We only have three to four days to build another canal.'"

"It totally changed the meeting," Terry said. "People believed in him and believed him when he said they would be taken care of. There was still a lot of concern, but everybody knew it would be done."

"I remember we were saying no until President Gerratt stood up," said Willes Blauer. "He said that if everyone would agree to let it come through their places before we left the meeting, we would be fully compensated. That was good enough for me. I remember Val Robins stood up and said he'd go along. I followed him. I said I had wanted to be the first one up, but Val had beat me to it. Still, it was hard, letting them come through."

"Rex Gerratt was the inspiration behind it," said Lane Gochnour. "It never would have been done without him. At times when we'd get to arguing and scrapping, he'd say something—I can't remember exactly what—something about we needed to get it done. I wish I could remember his exact words."

Marvin Blacker later remembered that while Rex was talking, he was thinking, "I wish that man would go home."

"I could fight the county commissioners," Marvin said, "but when the stake president stands up, you don't have any defense."

Ernie Straubhaar left the meeting. "I got my foreman lined out and began calling for other equipment," he said.

Roy Beams left the meeting and called Salt Lake. The Armco shops there would need to make the big barrels for the BID siphons. It was an emergency, he told them. They needed to begin immediately.

Evan Rasmussen remembered how his feelings changed. "As the meeting progressed, I was becoming more and more impressed with the people of Cassia County, with their willingness to do what needed to be done. By the time the meeting was over, I thought we'd lick the problem."

"The meeting was the turning point," said Terry. "It marked total commitment to the project. It was inspirational. It made it successful. After the meeting Tom Tate was committed."

"I walked away from the meeting feeling that something had happened," said Leman Messley. "I remember thinking, it's got promise. It was so different from the feeling I'd had at the earlier meeting. People came out with a sense of direction. There were job descriptions, responsibilities assigned."

Leman remembered some literature that he had read about grief counseling. It seemed to him that the whole community was going through a grief experience together. The fear, anger, denial, helplessness—all the emotions that one would expect to see—were there. The whole community was going through what any member might be expected to go through individually. It was a healthy process going on. Like a person working through these emotions, the community was coming out stronger. They were able to deal with the facts.

Tom Tate called the Corps office in Walla Walla. Under ordinary circumstances he would have asked for the Corps' participation in preventative measures. But permission for flood prevention had to come from Washington, D.C., and it would take too long. The district office in Walla Walla could authorize a flood fight, but only when a flood was in progress. Tom requested permission for a flood fight in advance. Though unorthodox, it was the only way.

Terry Bingham remembered Tom saying to his bosses in Walla Walla, "'I need authorization for a flood fight.'

"'Time?'

"'Three days.'

"'Can you do it in three days?'

"'No, we need cold weather also.'

"'Will it be cold?'

"'Can't say. These people are going to do it. It would not look good if the Corps stood by and refused to help.'"

Tom's support of the project and his reputation were crucial in getting the Corps' help, Terry remembered. Before receiving authorization, Tom was calling for machinery. The contractors moved equipment at his word, without waiting for an official agreement.

Tom was sure they could complete the project. "I wouldn't have started it if I didn't think we could get it done in time," he later said. "My boss asked if I could do it. I said yes. He left it up to me. I was not sure we could get to the river, but I knew we could get to the draw. We might lose a few houses out there, but it was better than going through Burley. Terry and I talked and said, 'We've got to try.'"

County attorney Al Barrus did not think about whether it would succeed. "We had no choice," he said. "Our whole effort was to do it. My gut feeling was that it had to work. Such great people were involved."

Burley Stake President Boyd Poulton worried about those whose land would be affected. "My biggest concern was what we were going to do to make it right with the farmers. I was raised on a dairy farm. I knew the problems it would cause them. I was concerned about their lives being destroyed or disrupted."

Glen Jensen was apprehensive. "With a new canal you don't know the erosion problems," he said. "If the canal silted in and broke out it would do tremendous damage. I was afraid we would stir the countryside and not get the job done."

"People were coming into my office, crying because they said it was too late," Rex Gerratt remembered.

The Cassia County commissioners, lawyers, Burley Mayor Chuck Shadduck, and city council members headed for the mayor's conference room. The decision had been made. Now they had to make it legal.

"The county wanted to obtain easements," city attorney Bill Parsons said. "In our opinion there was no time or money for easements."

Although most landowners had signed blank easements at the meeting, a few had not signed. It would take time to get those signatures, time to find the landowners who had not attended the meeting or were out of town. It would take time to get official agreement from the railroad; it was questionable as to whether the railroad would grant permission. Yet the county could not cut through property illegally; the Corps would not move one shovelful of earth until all permissions were obtained.

They also had to figure out how to pay for the canal. The commissioners were not authorized to go into debt or spend money outside their budget. Yet SCS officials were surveying the route and Ernie Straubhaar was calling for men and equipment. They had to do something to make it legal.

County Commissioners John Adams, Norman Dayley, and J. Weldon Beck on the steps of the Cassia County Courthouse in 1989. (Photo courtesy of Ryan Thomas.)

All the lawyers were vaguely aware of an old law, a public purpose doctrine that had not been used in Idaho since 1910. Under this law a government entity, such as a city or county, could invoke emergency police powers. They could take a person's property, use it, damage it, and return it with no compensation to the owners, if necessary to protect the public. The law had been used mostly in the 1800s in cases of fire. A town could pull down someone's house or business to keep the rest of the houses from burning.

"It was the only feasible way to do it," said Bill Parsons. "The best evidence that we had from the engineers showed that there would have been great destruction. It was necessary to damage some people's property so that the rest would be saved."

At 3:00 P.M. the commissioners informed the press that they were building the canal under the police powers act, and that the landowners would receive no compensation. They sent word to Ernie Straubhaar not to begin moving dirt until the legal papers were signed.

Unaware of the county's decision, several men were asking for easements. "They had printed up a sheet for each farmer to sign," said Tom Geary. "Some of us started getting signatures. Harold Bowers and I worked together starting in the Oakley area. I don't remember anyone who didn't sign."

"David Beck and I were assigned to contact all landowners from Golden Valley Road to the river," said Grant Wyatt. "We got a signed statement from them. We had no problems—never had one refusal. We assured them their loss would be compensated. We couldn't find some absentee owners. We didn't know what we were going to do about that."

Bob Sagers talked to landowners along the south third of the canal. The only signatures he did not get were from people living out of state, one in New York, the other in California.

Oakley farmer John Martin, who had not attended the meeting, signed an easement. "Of course you don't like it," he said. "But people have to do what they can. They can't just think about themselves; they have to think about the disaster as a whole, what's best for the whole community."

"In the back of your mind you wondered what it would do to you financially," said Randy Hardy, who farmed on both sides of the railroad track, "but I also knew it was the best way. The old Goose Creek channel wound around so much. So many homes were built on it and by it. It had to be done, so we climbed in there and worked with them, helped them to destroy our property."

The Star area, from 200 South to the river, was part of the Burley West Stake. A representative from that stake talked to the landowners and asked them to sign an agreement, remembered President Walter Petersen.

"It was hard to ask them," he said. "It was a true sacrifice, a potential sacrifice of dollars and property and homes. We'd explain that we'd try to minimize the loss. We would sandbag and dike around homes, but no one could guarantee that their homes would not go. In most cases the response was, 'We're scared, frightened, we don't know what it's going to do to our land or our home. We don't like it, but let's get it done.' Only one or two were opposed, the ones whose homes were closest to the channel, but even they signed the agreement.

"Many tears were shed with the landowners," he said. "I remember thinking, 'What great people, what unselfish people.' They would have chosen not to do it, but were willing because of the overall picture."

Although no one from Burley Stake lived along the route, Stake President Boyd Poulton, who knew the farmers because of his real estate business, also helped obtain signatures.

"It was the most difficult assignment I ever had to do in my life," he said. "I'd grown up on a farm. I knew how hard it was for them to think a canal would come across their field. I do not recall anyone being angry—perhaps there was some anger, but I didn't see it. I saw a lot of fear, a lot of concern, but no anger.

"I remember talking to one family. They were hard workers, of limited means. I remember telling them, 'It's going to be all right.'"

Fred Hawker remembered seeing some of the farmers after the meeting. "You don't often see a grown man cry," he said. He remembered everyone telling him, "You're in the natural channel. It won't hurt you so bad."

Her husband and father-in-law were upset when they arrived home after the meeting, remembered Julie Randall. "They're going to ruin us," Grant Randall said.

"Not too many people were for it, but what could you do?" said Julie. "Dave Beck was not thrilled. Steven's dad was not thrilled. They went along. They liked President Gerratt. He made it possible for them to work through the problem. He didn't like to do it either, but he had to. They liked and respected him, and figured he would do what was best. And so they went along."

"Right at first you didn't want it to happen," said David Beck. "But it was just something that had to be done."

As the afternoon wore on, farmers began hearing about the police powers act and learning that their property would be taken without compensation.

"People were calling," remembered Steve Bywater. "They were saying, 'They told us we would be compensated, and now they're saying on the radio we won't be. What's happening?'

"I couldn't give satisfying answers," said Steve. "I could only say, 'The county can't compensate you, but we hope you'll be compensated in some other way.' It was not a fun time."

The police powers act was a blow to the landowners. “The police order was unnecessary and offensive,” said Lane Gochnour. “If someone takes something away from you, it’s not the same as you volunteering.”

“The police power—that was the worst thing,” Willes Blauer said. “We’d volunteered—we’d said they could come through our land—then they went ahead and took it from us anyway.”

“I don’t think any of us were calm,” said his wife Norene. “To have a piece of ground that you think is yours—the county comes in and says it’s condemned—it’s not yours any more—that’s real frustrating. When you’re looking forward it’s hard—it’s the unknown that’s hard—we didn’t know how long it was going to stay there. I was real depressed for awhile.”

Marvin Blacker feared that he would lose his home and farm, “but that wasn’t the hardest part,” he said. “The hardest part was living in America and thinking we had all these freedoms, and then finding out we didn’t have any. They could go into and take your home and do anything with it—they could turn it into a morgue if they wanted to—without any compensation—if it was deemed a necessity. It didn’t matter what damage they did to it. They didn’t owe us a thing.”

Some of Marvin’s neighbors, whose place would not be affected, told him they would help him block the canal. “They would have planted themselves in front of the tractors like they’d seen people do on television,” he said. “I told them not to do it. I could have stopped the canal. I could have hired a lawyer in Twin Falls and got an injunction. I didn’t have to stop it permanently. All I had to do was stall for twenty-four hours and nature would have taken its course.”

Marvin did not call a lawyer. “We didn’t like it, but we didn’t try to stop it,” he said.

“We couldn’t be the bad guys,” said Marvin’s wife, Lovel. “We couldn’t be the ones to stop the canal and flood the city of Burley. It’s our community. We live here too.”

“It was a real test,” said Marvin, “to see whether you’d follow what your church leaders wanted you to do. I think it was a test for everyone. We each failed in our own way.”

At attorney Bill Parsons' office, city and county lawyers frantically tried to complete the paperwork to make the building of the canal legal. "We had to dig out cases that supported our position," said Bill. "We had to make sure the state had not preempted the right of the commissioners to do this. We had to draft the ordinance and do all the other legal hocus pocus. It had to be done according to due process. We had to make sure they were doing it legally."

At 11:10 P.M. the county commissioners went to the court house to hold an official commissioners' meeting.

"The board met this date with all members present with County Attorney Alfred E. Barrus, Deputy Attorneys Steve Bywater and Howard Smyser present," wrote county clerk Frank Kearns in the official minutes. "This special Board meeting was held to consider the passage of an emergency ordinance regarding the consideration of a flood canal from the Oakley Dam to the Snake River.

"After due consideration and review of all information provided to the Commissioners by the A.S.C.S. office, the Corps of Army Engineers, private contractors and other sources, it was moved, seconded and made unanimous that Ordinance 84–1 be adopted as follows:—"

Though the official minutes then quoted the text of the ordinance, at 11:10 P.M. the document was not ready for the commissioners to sign. They went to Bill Parsons' law office and dozed in the lobby while they waited for the lawyers to finish. Weldon Beck went to sleep on the floor. It was three or four in the morning before the lawyers completed their work. The commissioners took the ordinance, returned to the court house to sign it, and adjourned their meeting. Construction could begin.

Digging the new canal.
(Photo courtesy of Donald J. Clark.)

Chapter Nine

Thursday, May 17

At dawn Thursday a hundred pieces of equipment waited near the elevators west of Oakley, ready to dig the canal. "We were going to start Wednesday night," said loader operator Lawrence Burch. "Then they decided to wait until morning. A guy in a tractor drove straight through a field and said, 'Follow me.' That was how they marked our way. They told us, 'When nothing is staked, follow the railroad tracks.'"

They leapfrogged equipment, Lawrence remembered. "I had to drive three miles to find me a spot to dig at first."

Lawrence had never seen so much equipment working together. "It made the goose bumps stand up," he said. He radioed for a photographer to fly over the site and take pictures.

But though Lawrence was excited to see the equipment, he was not optimistic about the project's success. "All the operators were saying it would never work," he said. "They were saying, 'It's too far to go. The water's rising too fast.' I wasn't saying much, but I had my doubts. Down deep I didn't think we'd make it. But every one of us was bound and determined to stay there and dig until the water pushed us out."

Many construction companies volunteered equipment. "They were ready to donate machinery and time," said Kloepfer office manager Blaine Allphin. "When you get in a crisis everybody wants to jump in and help."

Equipment being hauled to the construction site near Golden Valley Road. (Photo courtesy of Jim Mai.)

A lot of people called construction companies, Rex Gerratt remembered. "Terry was calling, Tate was calling. Tate said, 'I want so many cats and they have to be of this size.' We needed big equipment—not many people have D–9 cats. They were pulling equipment off other construction projects to come.

"I called our church regional representative Martin Zachreson who was a businessman in Boise, and said, 'Can you contact some construction companies up there?' They sent carryalls, loaders, and cats. We didn't want anything smaller than a four-yard bucket loader."

"President Gerratt called us, told us what equipment they needed," said Paul Stake President John Stevenson. "We'd get on the telephone. We called anyone who had equipment, told them it was volunteer service and asked if they would be willing to help. We were never turned down. When people see there's danger, they respond.

"I remember going into the bar at Paul. Howland Croft, who owned the bar, had two large Caterpillars that we needed. When I told him what I wanted, he said, 'You'll have to give me a ride to my place so I can get them for you.'"

Ernie Straubhaar directed construction from the dam to Golden Valley, the Corps of Engineers from Golden Valley to the natural drain. Frontloaders and carryalls dug the first eight or nine miles while the cats started at Golden Valley. "We had contracted materials and equipment mixed in with the volunteer equipment," Terry Bingham said.

Equipment lined up for fueling and servicing. (Photo courtesy of Leroy Jarolimek.)

They planned to construct a canal seventy-five feet wide, three feet deep, with banks three feet high and twenty feet wide.

Someone asked Tom Tate whether the canal would work, Terry remembered. "Tom said the Corps hadn't built too many canals, but they would do their best here. This was probably the longest canal they had worked on, but the other one was the Panama Canal and it had been successful."

A story began circulating: They were building such a big canal because the only other canal the Corps had built was the Panama Canal. The Corps did not know how to build anything smaller.

On Wednesday afternoon Kent Foster had made a preliminary survey to determine the route for the canal. "The first part was a mixed up area to go through," Kent said. "We surveyed and set flags across a field. It was only preliminary—we had not definitely decided that the route would go that way. During the night someone moved the flags and when we came back in the morning, the canal was dug."

The commissioners were angry, Kent remembered. "They asked me to survey and see if the new route would work all right. It would. It was a good route, maybe the best one, maybe the one we would have decided on. But it made people upset. If we could have waited, we would have got people together and discussed it, would have talked to the landowners. After that the commissioners told me that the SCS had the authority to pick the route, and nobody was to change it from what we decided."

"Al Barrus, the prosecuting attorney, straightened people out," said John Adams. "He said, 'If you dig where it isn't marked, you're in jail.'"

"There was some difugalty," remembered Larry Adams. "If you had your druthers, you'd rather the canal be in the other person's field. There were a lot of people there, and they were trying to decide where to take it. Somebody would say, 'Go on this side.'

"Another person would say, 'No, that's my field, go the other side!'

"It was funny to watch. An hour later they were friends again, but for awhile they were ready to take a baseball bat to each other."

"One big problem was that too many chiefs were trying to tell Ernie what to do," said attorney Steve Bywater. "People were saying, 'You need to direct the canal around this thing. You can't cut this mainline.' The resolution said Terry was to supervise. Ernie was to take orders only from Terry. Once that was decided, there was no more trouble. I spent a lot of time talking to farmers, telling them that it wasn't what they said we could do, it was what we were going to do."

"Weldon Beck asked us to run a line from Oakley to the dam," said surveyor Steve Pearson. "We felt it should run on a straight line no matter who it affected because water flows better in a straight course. We put the first wheel tracks down through there. It was a field of spuds—they'd been planted but were not up yet. I felt so bad about that. Having been born on a farm, I have a hard time driving over crops that have just been put in the ground."

Ernie Straubhaar met at the job site with the commissioners and lawyers to decide what they would try to save for people. "We decided we couldn't salvage anything," he said. "We had to get her dug. There was no time for anything else."

"There was really a scurry ahead of you," said equipment operator Lawrence Burch. "They told us, 'People have been alerted. If anything is in the way, haystacks or anything, doze them down.' You could see people hooking with pickups, tractors, whatever they could use to get stuff out of the way. The same with water lines, telephone lines. We wouldn't know we was covering them up until we'd see a wire sticking up. By the time you see them, it's too late."

Lawrence covered an old combine. "I didn't move it, just covered it where the bank was. It was probably the last priority for someone to move."

Some high school students helped Randy Hardy move his fence. "They barely got it out in time," Richard Hardy remembered. "We went along pulling out fences as fast as they dug."

"We never saw our backhoe too often," said Kim Cranney. "We took it out and left it for anybody to drive. It was too small for digging the canal, but people could use it to dig out mainlines or whatever. A neighbor hired a man to operate it." Other farmers left their trucks by the canal with the keys in them. They were there for anyone to use who needed a truck.

"My trucks are out there somewhere," a farmer later told Bill Matthews. "I don't know where they are or who's driving them, but they must be doing something because they fill up at my gas tanks each night."

But at Golden Valley road, the cats sat idle as operators waited for instructions from Corps officials. Since Wednesday afternoon, Corps hydrologists had been trying to estimate the runoff. Denny Davis fed information about local conditions into the Bureau of Reclamation computer in Burley. Engineers in Boise, Walla Walla, and Portland analyzed the information on their computers. According to the USGS gauge, on Tuesday 1,320 cfs flowed into the reservoir from Goose Creek; but by Wednesday the flow had dropped to 870 cfs—an indication that the peak had passed.

"Sweet relief," said Terry Bingham when he heard the news.

"Flood Threat Declines," the *South Idaho Press* announced in its headline the next day.

Corps engineers, who had initially used Tuesday's data to determine the size of the canal, recalculated using Wednesday's figures. The construction operators waited.

"It seemed so slow at that point when they couldn't get started," said Terry. "But Tom Tate had everything all lined up waiting for the signal."

Watching the idle equipment was more than Bill Cranney could stand.

"One cat was stopped waiting for someone to give the word," Bill remembered. "I said, 'Dig!'

"He said, 'Who are you?'

"I said, 'I'm the boss.'

"Later someone else came up and said, 'Who told that cat to dig?'

"I said, 'I wouldn't know.'"

But while construction operators waited and engineers made new calculations, observers at the dam reported that a huge amount of water was entering the reservoir. As people watched their sticks in the water and their marks on the spillway, they could see the water rising at an incredible rate—more than an inch an hour. The water was spread over the entire surface of the reservoir, yet rising faster than it had ever risen before.

At 2:00 P.M. Thursday afternoon the engineers completed their calculations. They sent word that the canal should be forty feet wide at the top—about half the size originally planned. The cats rolled off trucks and began digging.

Mike Cranney, who was at the construction site, was dismayed to see the size of the canal. He protested to Corps supervisor Donna Street. "This is what the computers have figured out, and this is what we're going to do," she said.

Mike called Rex Gerratt. "I grabbed Wade Baker and called for people to meet in town," Rex said.

Tom Tate, Terry Bingham, Weldon Beck, Rex, Wade, Mike—about fifteen people altogether—met at the city hall.

"We told them, 'Your canal is not wide enough,'" Rex said. They talked a little, discussing the differences between the gauge readings and reports from people watching the reservoir. "When our seventy-five-foot canal hits your forty-foot ditch, someone is going to look pretty silly," Rex said.

"It didn't take long to convince Tom," Rex later remembered. "It was decided in about a half-hour's time. Tom talked a little and then said, 'Well, President Gerratt, if you feel that strongly about it, we'll stretch it out to seventy feet.'

"It was one of the miracles," Rex later reflected. "There were all the computers and all that scientific equipment, and Tom was willing to go against all that, willing to listen to a bunch of farmers—that's what most of us were. I think it was because he had worked with us so long. He'd gotten to know us, knew what we could do."

At 3:00 P.M. the Corps began digging a seventy-foot canal. Denny Davis, who had been keying information on his computer and watching the engineers' calculations, heard about the change in plans. "I went to Leo Busch and said, 'They're making the canal, but twice as big as what we told them.'" Everyone was incredulous, he remembered. "We all thought they were crazy."

Tom asked Bureau of Reclamation official Earl Corless to monitor the reservoir's increase. Ordinarily a hand-operated gauge on the tower was used to measure reservoir increase. Earl had a pen-and-ink device in a barrel that gave a continuous reading of the water's height. "We wanted to get a curve drawn so we knew how fast the water was coming up," said Denny Davis. "We could tell by the slope, by how steep it was, how fast the water was rising."

Corps hydrologist Bob Rickel in Walla Walla asked the USGS officials to check their gauges. Thursday evening Joe Bunt discovered that someone had pinched the gas line on the Goose Creek gauge causing it to malfunction. It had been giving incorrect readings since Tuesday.

At the construction site more cats and loaders arrived. Donald Clark estimated that about 250 pieces of equipment were working on Thursday. Delma Bean watched equipment rumble past her house. "Whenever we'd see a big rig going down to the canal, we'd say thank you," she said.

Terry Bingham's notes became increasingly sketchy as he tried to keep track of everything. "The next days are a blur of running from one end of the project to the other, meeting with property owners, monitoring three multi-frequency radios, negotiating streamflows between Dave Hollingshead and Lee Baker and hoping the cool weather would hold," he told the *South Idaho Press.*

The city council members decided they needed a liaison between the city of Burley and Terry Bingham. "We wanted to keep track of what was going on so we'd know what we'd have to do," said Ken Fronk. "I volunteered."

Terry was upset. "I have nothing against Ken Fronk," he said. "He's a nice guy. But I was miffed. Here we'd been working for weeks to save the city, and then at the last minute, they send somebody out to ride with me. I thought they didn't trust me."

But it worked out all right, he remembered. "It was getting so hectic. Ken helped me with phone calls and ran the radio."

"I tried to help where I could," said Ken. "I kept notes on who was going where and doing what. I don't think Terry could have managed by himself. He was keeping track of what was going on and coordinating efforts. If somebody needed something—food, fuel, equipment, trucks—we'd try to get them to people. We were in a pickup, going back and forth."

Since the project had begun, people used two-way radios to communicate. "I was answering phones, the radio—we had a whole bank," said Thelma Clark, secretary for Cranney farms. Their office was headquarters for the Oakley Highway Department.

Many farmers had radio systems, but only a few systems were linked. Each farmer's home or office became a relay center, with someone taking the radio message and calling on the telephone to another person who then relayed the message by radio to the person on the canal. "We learned real quick how to contact somebody who could contact somebody," said Bruce Beck.

"We bought some CB radios when the emergency started," said Wayne Mullen. "Luther Gardner, our ward emergency director, was my right-hand man. We had a radio for his pickup and one for my pickup. Our base station was at the house. If I needed to talk to someone, I'd call home and Elda would call that person on the telephone."

"My daughter Carolyn and I were on the radio and telephone constantly," said Rex Gerratt's wife Marjorie. "I could not take five steps without the phone or two-way ringing."

They needed better communication. Terry Bingham knew that Boise Interagency Fire Center (BIFC), which handled emergency communications for the entire United States, had radios available for emergency use. Harold West, electronic technician with the Bureau of Land Management, could request them.

"Terry came to see me," Harold said. "I wasn't surprised. We had been friends for years. Terry had always said, 'If we have a disaster, I'm going to requisition you and your family for communications.'"

Harold called the district manager in Boise while attorney Al Barrus filled out papers. "We talked about it one day," Harold said. "The next day we had the radios. Everybody from the state was really helping us. They knew about the emergency down here."

Jerry Rauscher brought the radios from Boise and helped Harold install a repeater on a ridge so that everyone could talk with everyone else. Jerry brought seven radio kits with 16 radios in each—112 radios altogether. "They were Motorola MX 360 radios," said Harold. "They were real expensive right then, cost three thousand dollars each. They were the top brand, the top of the line."

It surprised BLM employee Ken Fuller to see Harold distribute the radios so freely. Harold was responsible if any were lost or damaged. "He had to trust people," said Ken. "What other choice was there? Harold is a local—knew the people he was dealing with."

Harold was not worried. "I gave them to whoever needed them," he said. "I had no qualms, no concern about not getting them back. I was concerned to get the equipment out and get it going. I gave radios out a little too fast at first. I didn't write down who got them. There was a panic to get everything going because of the flood. You do things fast and then sort it out later, the same as with a fire."

He monitored the radio conversation. "I'd listen for reception rather than the message," he said. "If someone said their radio wasn't working well, I'd take one out and exchange with them. I'd repair theirs, recharge the battery. To help us save money, BIFC took batteries that they had saved from old kits and gave them to us free. They always replaced batteries with new ones even though the old batteries might still be good. You can't have a battery go out when you're on a fire. The batteries cost about seven dollars apiece, and we replaced them every other day."

The Nelson, Sagers, Anderson, and King accounting office became the message center. "Someone was there around the clock," said accountant Newel Nelson. "Our secretaries took care of it during the day. At night couples came in—worked four- to six-hour shifts. We had a regular schedule as to who would be there and what time. We had a list of people and what radio number they had.

"Three counselors from three different stake presidencies were part of our office. We concentrated on the flood the whole time. We didn't do enough office work to shake a leg at."

Corps supervisor Larry Johnson talking on the radio while other workers enjoy a lunch break. (Photo courtesy of the Walla Walla District, U.S. Army Corps of Engineers.)

"I was off work for two months," Oakley Stake counselor Bob Sagers said. "I'd check in about every morning to see if there were messages. Nobody complained. Everybody understood."

Police Chief Leman Messley organized a flood emergency hotline. "The telephone company gave us five or six phones," he said. "We set them up in the civil defense room in the law enforcement building. We had volunteers answering questions around the clock."

"I was in charge of lining up people," said police secretary Joan Easton. "I'd work my regular eight hours, run home, grab a sandwich, and be back until midnight. We took care of the jails and went to court. All the rest of the time we worked with the flood."

Most questions to the hotline came from farmers, remembered Inez Messley, one of the volunteers. "They had legitimate concerns. They wanted an hour-by-hour report of the progress of the construction, wanted to know how quickly the water would reach their place."

"It was hard work," Joan remembered. "There was so much emotion. Some people were so angry." But she also remembered how much support they had, "how everybody grew together during that time."

The hotline helped to quell rumors. Some people heard that the dam had broken and called to ask how quickly the water would arrive. Later the police learned that a young man dressed in an old national guard uniform had gone house to house telling people to evacuate.

"My son called," Willes Blauer remembered. "He said, 'What do we do?' I told him to call the police. I don't think the kid was doing it as a prank. I think he thought he was really helping. He was a little bit off."

"There have been crazy rumors again today," wrote Karlene Hardy in her journal. "I'm learning not to believe everything I hear."

People were calling the commissioners, remembered Weldon Beck. "I couldn't sleep, couldn't eat, was going all the time. As far as the digging was concerned, we hired Ernie and left it to him. But if Terry needed us or the supervisors needed us to see somebody a little irate, they called us. Somebody was concerned about water in their basement, somebody saw some chickens floating, some calves were without feed. That's how you learn to know people, when you work with them under pressure."

He told everybody that he was not running for county commissioner again. The emergency was too much for him—the county needed a younger man.

"Dad was on both sides of the fence," said his son Bruce Beck. "He was a farmer and on the committee to make the decision. He was making a decision that could affect people's lives for the rest of their lives. I remember watching him age."

The SCS designed and surveyed the route. "We did not survey everywhere," said Kent Foster. "The route and design was done off maps except for a few problem areas. I had crews of guys working, BLM guys working on the lower end from the railroad tracks to Hawkers."

Area engineer Mark Jensen, area conservationist Joe Icenhower, and hydraulic engineer Lonny Fox came from the state SCS office to help; Rich Lavender from the state office had been helping with the surveying for a couple of weeks.

Burke Scholer surveyed the route between the railroad and the natural drain. "We started Wednesday after the decision was made," he said. "We were flagging the route, checking every hundred feet. I did all the design work off survey notes. It was a helter-skelter affair. I ran the grade in my head as I went through with the survey. I had the design done about the same time the survey was complete. There was a problem in Marvin Blacker's yard, a little bottleneck-type thing by his corrals. Once we got past there, it was clear sailing."

"Things went smooth for something that big," said Kent. "Damage-wise, some people were hurt more than others, but very few people were griping or protesting. People didn't lose their heads and start shooting someone."

Although BLM range conservationist Ken Fuller ordinarily did not do much surveying, he and archeologist Isaac Barner surveyed along with the others. "Whoever could do it, they got," said Ken.

He drove his truck in the field so he could take radio calls. "I'd run back and forth from surveying to the radio," he said. "Northwest Pipeline out of Salt Lake called. They were concerned about the construction route. Idaho Power was concerned about their substation. Undersheriff Billy Crystal worked with a lot of panicked people. I saw a lot of him because I was doing the lead survey. He was out there when there were problems—get a sprinkler pipe out of the way, how wide and how deep should the channel go here, how can we control and contain damage. All agencies worked together well. Once a structure is in place as to who calls the shots and who helps, you can get a lot done in a hurry."

J.J. McLaws surveyed the grade, making sure the canal would not have high spots that would impede the flow of the water. "Several people who saw us thought we was determining the route with the cats practically on top of us," J.J. said. "But we had to work right along with the cats to show them how much to dig."

J.J. was surveying when Bob Sagers arrived with a pickup filled with sandwiches. "Somebody said, 'There's President Sagers,'" J.J. later remembered.

"He came up to me, and I said, 'Where do I get a title?'

"He said, 'What do you mean?'

"I said, 'I keep hearing, President Sagers, President Gerratt, President Petersen, President this, President that. When anybody wants to talk to me, they just say, "Hey, you." So what I'd like to know is where I can get me a title?'

"He just grinned and said, 'Well, about the best thing I can give you is a sandwich.'

"Every religion worked together well," said J.J. "This is the best neighborhood in the world."

When it was time to eat, somebody was there with a sack lunch, Lawrence Burch remembered. "They'd toss it up at you. You'd eat on the go, dirt and all." The Salvation Army also brought food, he remembered. "They had vehicles with drinks, sandwiches, pastries, everything donated. But you had to go to them. They didn't come to you."

Someone brought Earl Corless a sandwich while he was at the reservoir. "It may seem like a little thing," he later said, "but I remember that. I remember being hot and hungry, and someone bringing me a nice lunch with a sandwich and cookies and oranges. Sometimes it was the Oakley people, sometimes the Salvation Army. People may think it's no big deal, but it was for me. I can't go past a Salvation Army kettle without thinking about that sandwich, and dropping something in."

"The man from the Salvation Army was here night and day," police secretary Joan Easton said. "He'd come in with sandwiches. He'd say, 'Where do you want me to take them? Where's the crew?'"

The two Burley stakes made sandwiches for people on the north end of the channel. Cloyd Taylor, owner of Price's Cafe, offered to help. "We donated lunches and took them where they wanted them to go," he said. "The Relief Society helped us put them together. We set up space in our back room, had everything ready by the time people got there. It was easier to do it at the cafe than to pack stuff somewhere else and set up."

Rachel Gorringe, Burley West Stake Relief Society President, assigned the wards to make sandwiches at the church also. "The first ward I called was the 11th Ward," she said. "I asked Edith Dille to be in charge because she had been in charge of the school lunch program for years and had the expertise. The women charged supplies at the stores and the stake paid for it."

Golden Wood with a box of lunches for the canal workers. (Photo courtesy of Rachel Gorringe.)

Evan Rasmussen and crew worked on the BID lift canals. "We learned from our experience with the Milner canal," said Evan. "We built siphons to take our own water under the new canal. We knew how much water our canals carried, could control that water."

They needed culverts for the first- and third-lift canals, the G and J. The H canal, which ran under the 200 West road, already had a culvert. "We installed a 72-inch siphon for the J canal, a 60-inch one for the G canal," said Evan. "We ordered large siphon barrels from Armco out of Salt Lake. They had to be special built. We dried up the canal, excavated the trenches. The siphons were delivered at 2:00 A.M. Friday morning. We installed the siphons, poured the head walls and collar around the joints to prevent leakage. We were really concerned about getting done in time. I didn't take my boots off for four days."

Farmers along the route rearranged their irrigation equipment. "We had to plumb pipe lines and get our irrigation system tied together," said Doug Cranney. "We took pipes and electrical wires either under or over the canal. They went through with the digging, then we could start our work."

"My neighbor just up the road, Gerald Lind, was a retired engineer," said Randy Hardy. "He helped me figure out a way to take the water across the canal. We had an eight-inch pipe to carry

Digging an irrigation pipeline at Beans' farm. (Photo courtesy of Nona Jenks.)

water inside a ten-inch pipe for stability. Wayne Mullen and Luther Gardner did the labor."

"The canal cut through our pipe lines in three places," said Roland Bean. "We made trestles with railroad irons to put the pipe over the canal."

"We had to build a new dam," Fred Hawker said, "had to move the pump and span the channel with pipe so we could irrigate on both sides. We had to reroute the electricity. Chuck Jones and Grant Saxton of Unity Light and Power came out and did that gratis for us, even though I'm under Idaho Power. Grant Wyatt and other neighbors helped. We borrowed pipe from the neighbors."

"We sandbagged our house, sandbagged our well, sandbagged everything," said Lovel Blacker. "We took everything out of our basement and prepared to move at a moment's notice. We had a lot of help."

"It was well organized," said Marvin. "Most organizing was done through the church, though it was our neighbors who were not church members who gave us some of the best help."

People brought food, Lovel remembered. "Binghams brought us some stuffed potatoes. President Gerratt came by and we gave him some. He sure went down on those potatoes. I don't think he'd had anything to eat all day."

Bill Roberts sandbagging at Blackers. (Photo courtesy of Lovel Blacker.)

"Here we were," said Rex Gerratt, "cutting through the middle of Marvin Blacker's place. Marvin didn't know whether he'd have a house or farm left. A lot of guys would have come at you with a shotgun. I went there and he fed me dinner."

Kent Foster and other SCS employees helped the farmers prepare. "We worked with guys on moving mainlines, worked to resolve little issues," said Kent. "We tried to minimize the damage, tried to make it fair, tried to make it as easy on everybody as we could.

"We had a problem with Freddy Hawker—the commissioners told us to talk to him. The lead surveyor had told him we were going to take out his irrigation pond. He was really upset. Burke Scholer and I went to see him and I said, 'What's the problem?'

"He said, 'They want to cut my irrigation pond, and I need two days to water my beets.'

"We told him no problem. 'We do have to take out your pond, but we don't have to take it right now.'

"He calmed right down then. We built all around him and left his pond until the very last thing—had a cat there to push it out when it was needed.

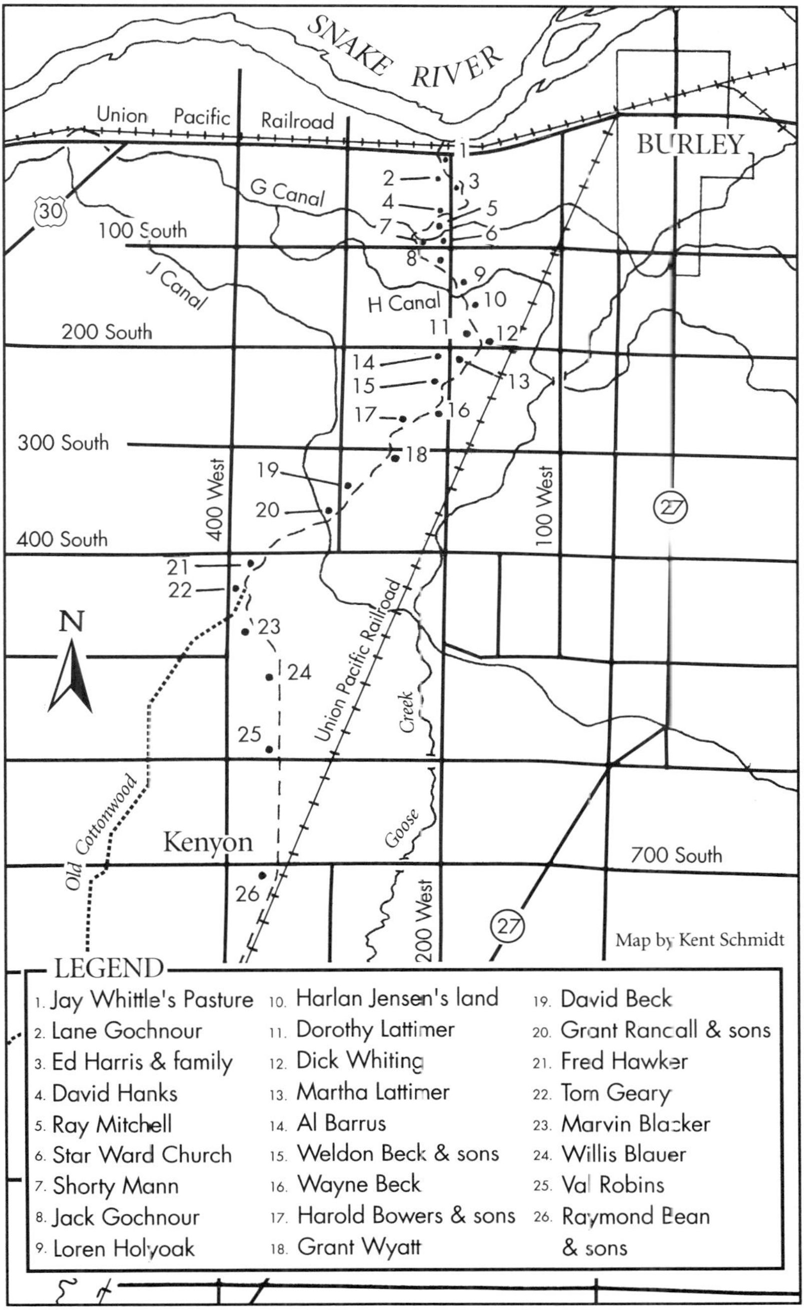
SNAKE RIVER
Union Pacific Railroad
BURLEY
G Canal
J Canal
H Canal
100 South
200 South
300 South
400 South
700 South
400 West
100 West
200 West
30
27
N
Union Pacific Railroad
Goose Creek
Old Cottonwood
Kenyon
Map by Kent Schmidt
LEGEND
1. Jay Whittle's Pasture
2. Lane Gochnour
3. Ed Harris & family
4. David Hanks
5. Ray Mitchell
6. Star Ward Church
7. Shorty Mann
8. Jack Gochnour
9. Loren Holyoak
10. Harlan Jensen's land
11. Dorothy Lattimer
12. Dick Whiting
13. Martha Lattimer
14. Al Barrus
15. Weldon Beck & sons
16. Wayne Beck
17. Harold Bowers & sons
18. Grant Wyatt
19. David Beck
20. Grant Randall & sons
21. Fred Hawker
22. Tom Geary
23. Marvin Blacker
24. Willis Blauer
25. Val Robins
26. Raymond Bean & sons

"We tried to accommodate, tried our best to limit losses. It's like fighting a war. You know there will be some casualties, but you do your best to limit them."

Tex Woolstenhulme, disaster coordinator for Burley Stake, had not been concerned about a flood until he attended the bi-regional stake presidency meeting on Wednesday night.

"They called us all together and said what they were going to do," Tex remembered. "I couldn't believe they were that concerned, wondering where to build the channel and how they were going to get it accomplished. It was mind boggling to me. I couldn't figure out what the sweat was about. I thought they could run water in the desert and let it soak away."

In spite of his feelings, Tex took care of the job. "I called for volunteers to help," he said. "We pulled out debris and fences, started at Highway 30 and went south. You do it whether you think it will happen or not. In situations like that, you've got to believe."

"People came and took down our fences," said Marian Harris. "We had barely put in fences, and they took them all down. We had livestock that had to be moved. We had milk cows—two or three—cattle and horses. My dad was living with us at the time. He was ninety-one years old. It was so traumatic for him. All this was land he had once owned."

"The wards were assigned to help people," said Burley West Stake President Walter Petersen. "In our stake the city wards were assigned to provide manpower and lunches."

"We had a meeting once a day with Bishop Melvin Call," said Bruce Beck, Pella Second Ward elders quorum president. "We rated the houses as to levels of potential damage. Bishop Call would say, 'Bruce, you've got an elder here, here, and here. I'd go talk to them. I talked to everybody who was not LDS who was in my age group. I'd say, 'What can we do to help you?' Then I'd organize people to get it done. We were trying to give the people encouragement. That was hard, because we didn't know what would happen or how bad it would be."

The new canal would come through some low places in Bruce's farm, but would not cause serious damage. He and his family also had ground near Goose Creek channel; a flood there would have damaged them more. He was not worried about his house, though his wife Dawn moved their scrapbooks out of the basement.

"It was tough for Dawn—I was never home," said Bruce. "We had nine-month-old twins, two other children besides. The oldest was four. I was working at least twenty hours a day—trying also to get my own farm work done. I remember a lot of nights thinking, 'If I can just have four hours of sleep, I'll be happy.'"

"May 17," wrote Julie Randall in her diary. "They worked real fast. When I got home from town, they were already tearing up things. By afternoon there was so many people helping on both sides of the road there were cars lined up. The elders quorum hauled our hay to this side of the road. Brother Baker dug a new water line for the barn. Steven had help with milking that night. There were lots of cars—sight-seers—looking to see what was going on. Steven said we should put a toll gate at the sign and charge because there were so many sight-seers. Bruce Beck, elders quorum president, told me to go to Farmers Corner to get pop and donuts for the crew. There were a lot of them."

"It happened so fast, and that was hard," said Rosanne Randall. "We were in shock, and still had to function, and fast. Plus we had to milk cows."

"The worst thing was the uncertainty, not knowing what was going to happen," Richard said.

"We had a lot of help," said Rosanne. "The ward moved the hay stack. Burley Stake moved our fences, moved and burned our wood pile. Lots of people made sandwiches. It was well organized. They assigned a town ward to look after us. They helped as much as the country people."

The canal, which came between their house and dairy barn, would also cut their private road that the milk truck used. "President Gerratt said, 'Bring your cows over to our barn, and we'll milk them for you,'" Richard remembered. "He'd have done it too. He's that kind of a guy. But, you've got to know cows. Cows don't like to be moved. We decided that if we could figure out a way to get the milk truck in, we'd leave them here."

The Lane Gochnours moved all their cows except the ones they were milking. "We moved all the cattle to Larsens," Lane said. "Moved everything down below up. Moved all the grain and all the hay. We hauled trailer loads of cows out. Somebody who hauls cattle into sale yards came out and helped us. We had to move our haystack, move a roller mill by hand that weighs one thousand pounds. The elders quorum came and helped us with that. We were worried. We didn't know what to expect. It's the unknown that you're afraid of."

Lane's wife Carla worried about their children. The youngest would be three in June. "I looked and he was with the calves when they were moving them," Carla said. "Another time I couldn't find him and he was in the grain. I could just see his little red jacket sticking up. Seeing my baby in the middle of that twice in one day was too much for me. I checked the kids out of school and sent them to stay with my sister in Soda Springs."

Though their house was high, "I had a bag with documents packed ready to go," Carla said. "The church passed out a paper telling us what to have in case we had to evacuate."

Val and Gloria Robins knew their house would be flooded if the canal broke. "We loaded everything that meant anything to us—pictures, scrapbooks—in my parents' motor home," Gloria said. "They took it to their place in Paul. We moved everything out of our basement. We didn't have any idea what would happen here. A side of the canal could have easily broken out."

Val Robins buried his four irrigation mainlines, which would be cut by the canal. He was worried about his potatoes. "We had eighty acres on the west side," he said. "It was a critical time. They were just coming up. I knew that if I didn't get water to them it would finish us. I couldn't get big enough pipe in Burley, had to order it from Twin Falls. We borrowed our neighbor's backhoe, then he needed it before the pipe came. We couldn't find anybody to come and weld the pipe. It was one frustration after another.

"Some people from over in Acequia brought us a loader—told us we could use it for whatever we wanted. I stayed up all night building a dike. Next morning the highway department said we couldn't do it, because it covered part of the road. They moved my dirt. After that everything I seen was red."

Some underground telephone cables ran underneath his place. The telephone company had marked them with stakes so they could move them. "In my anger, I took the stakes out and threw them away," Val said. "I don't know why I did that. I did and said a lot of things I'm not very proud of."

"I think everybody said and did things they later wished they hadn't, and that they wouldn't have done if they had not been under so much pressure," reflected Gloria.

"People get panicky and say all kinds of stuff," said Willes Blauer.

Gloria was Relief Society President for Pella Second Ward. She helped make sandwiches when it was their ward's turn. "I didn't think about it at the time, I was just doing my job," she said. "But later I remember thinking, here I am making sandwiches for the guys who are digging up our farm."

The people along the natural drain could choose whether to allow the water to come naturally or have the Corps dig a canal. "My place was the first one where they did not dig," remembered Bart Bowers. "I said, 'No canal. We'll let the water flow.'"

"We had them dig the channel," said Ed Harris. "We were afraid that if they didn't dig, the water would take our daughter's home."

The Randalls did not dig through their pasture. "The water would come close to our houses, but we are sitting high," Richard said. "I knew the natural channel would handle the water."

David Beck, who lived north of Randalls, was afraid that if he let the water flow naturally through his farm, it would back up and flood Randalls' dairy barn. "They left it up to me as to whether I wanted to dig or not," David said. "We needed to protect Randalls. We decided to let them doze out our place to save Randalls' dairy."

Although most people had turned their attention to the new canal, problems continued on the canal to Murtaugh. Jim Bright asked the Burley and Minidoka irrigation districts for help with the Milner canal. "The districts brought their equipment and machinery and took care of us at no cost to us," Jim said.

Tom Tate found Leland Baker on the Murtaugh canal shoring up a bank. "Tom said, 'Hey, it's all yours,'" Leland remembered.

"I said, 'What do you mean?' He told me that the Corps had been authorized to construct a new canal. He wanted me to be in charge of this one. He'd leave me some cats and operators.

"I said, 'What if I need anything?'

"He said, 'Call me, and I'll sign for it.'

"I think I'm probably the only person who had charge of the Corps, who didn't work for the Corps," Leland later reflected.

Tom came to the Murtaugh canal twice a day in a helicopter, Terry Bingham remembered. "Lee would tell Tom what he needed and Tom would get the equipment."

Containing the water was a constant battle. "I was there continually—couldn't leave for any reason," said Leland. "I was moving around grabbing backhoes wherever I could get them. In the daytime there was enough help to delegate and I could catnap. My brothers or neighbors would come and relieve me so sometimes I could go home and catch a little sleep. I'd go twenty-four to forty-eight hours before going home. I was not the only one—Kimber and a lot of them guys were doing the same thing."

Kent Searle remembered seeing Leland in the back seat of a car, his feet sticking out the window, taking a thirty-minute nap.

Whenever the water started to break out of the canal, Leland would need help in a hurry. Church leaders coordinated the volunteers. "I had five radios in my pickup," said Rex Gerratt. "One would ring—I wouldn't know which to answer. Someone would call and say we need this many volunteers. I'd radio home and call bishops and emergency directors. In a short time we'd have fifty or sixty people. People working at night could see the stream of lights coming out. This went on night after night.

"Lots of women were flagging trucks," he said. "They'd be out there for eight hours—left their little families at home. At night they'd be out with flashlights on the crossroads. They'd sit in their pickups until they saw lights, then they'd get out and motion the trucks on or through. They were posted at every intersection. Some had radios—could flag a truck down and give them a message. They were dumping a lot of rock to shore up the banks of the Murtaugh canal. The location kept changing as the water eroded the bank. There were about ten flaggers from where a truck left the road to get them to where they were dumping rock."

With most road crossings out and water flowing through the Murtaugh diversion canal, Adams Petroleum employees had difficulty fueling equipment. "The hardest part was where they'd want

Tom Tate and Leland Baker discussing the Murtaugh diversion canal. (Photo courtesy of the Walla Walla District, U.S. Army Corps of Engineers.)

you to put the truck," said Brett Adams. "They'd want you to go across water, go on the canal bank."

"They'd say, 'Ah, you can get through here,' when it would be a difficult place," Larry remembered.

Someone was operating a loader with a Bantam bucket at a place where Kelly Adams needed to cross the canal. Kelly climbed in the bucket with the gas hose and the loader swung him across.

Excavators were working on both sides of the channel when Bill Hinz arrived with a fuel truck, Bud Hoffbuhr remembered. "All Bill had to go on was a little road we had built on the canal bank. It was wet and miserable. He fueled one excavator but was stymied as to how to fuel the other. We said, 'No problem, just don't slip and fall.'" They arranged their equipment so that the two excavator buckets formed a bridge above the canal.

"Bill was not real excited about going across there at first," Bud said. "We kept the machinery well oiled. It was greasy and slick. It was a long ways to go, about forty feet. He had a big heavy hose that he was dragging. He walked across, but he walked very carefully."

Along the new canal, "a fuel truck would come out and park," said Lawrence Burch. "The equipment would go to the truck, suck it dry. The truck never moved. Whenever we fueled up we checked oil and water; otherwise we had no maintenance. If there was a breakdown in equipment, the repair crew was out there like an ambulance going to a wreck."

There were some accidents. Several people saw trucks tip over. "Slim Hewitt's dump truck dropped on Sam Lester's leg," said Kim Cranney. "We thought he wouldn't have a leg left. I ran him to the hospital. His leg wasn't even broken—only bruised."

At about two o'clock one morning Lane Gochnour was watching a backhoe break up the road by Ore-Ida when the backhoe tipped over. "All I could see was a cloud of dust and steam," said Lane. "I thought it was smoke. I ran there, thinking the guy was dead, hoping I'd be able to find some sign of life. When I got there, I could see that it was Paul Christensen, my neighbor, the best guy in the world. He got up, shook himself off, he wasn't even hurt. He'd probably been working three days with no sleep. We got a giant loader to upright his machine. He got back on it and kept working."

With so many men and women assisting on both canals and helping landowners prepare, the ones left at home worked long hours doing the routine daily tasks.

"It doesn't matter whose farm you're talking about," Leland Baker told the *South Idaho Press*, "for everybody who's been out here somebody had to double their efforts on the farm to make up for the ones that weren't there."

Kent Searle did not help with the construction on the Murtaugh canal. "I sent trucks up and my brothers went up," he said. "I covered for them. When I was up there, they covered for me. A lot of people covered for other people. Bill covered for Terry."

While Rex Gerratt worked full time on the flood, his son Alan looked after the dairy and his brother Larry took care of their farms.

"Rex usually helped out with the farming in the springtime when there was so much to do," Larry said. "I had to do a lot more running that year. I remember I was sitting with my hand on the number three pivot getting ready to start it. Rex called and said,

'We've got to have more help up here. We need more tractors to pack the banks. We're not getting enough response.'"

"I think I said something like, 'This is important too,'" remembered Larry, but he sent a tractor with a man to drive it. Larry drove the tractor at night.

Initially, Larry had been skeptical about the project. "My first thoughts were, 'It will never work. It's impossible to run water that far without flooding out people.' But that first night out there packing banks—you could see all the lights out there moving back and forth on both sides of the canal—at that point there was no doubt in my mind, I knew it would work. Nobody who was out there doubted that it would work. There was such a feeling to be out there taking part in such an enterprise."

At their farm south of Burley, Tom and Connie Geary could see most of the way to the reservoir. "It was so interesting at night," Connie said. "We would go out and look towards Oakley. It was like a Christmas tree—lights all the way, blinking and flashing around. We'd go out and look at that and marvel at what could be done when people put their energy and efforts together."

"It was devastating, yet exciting," said Tom. "I was so excited to see the cooperation and unity of the people working together. It was like in days gone by—people uniting in a tremendous project for the good of the community."

Reserve Deputy Kent Searle had been out of touch with the flood fight for a couple of days. Tuesday evening his wife had given birth to a baby girl at the Twin Falls hospital. "I came back from Twin Thursday night," Kent said. "I stopped at the sheriff's office. Terry said, 'Go home and get some sleep. It may be the last you'll get for awhile. We're building a canal.'"

Earl Corless walking along the spillway.
(Photo courtesy of Larry Adams.)

Chapter Ten

Friday & Saturday, May 18–19

"By now we were a well-oiled machine," said Terry Bingham. Each morning Tom Tate, Terry, the SCS officials, the commissioners, the church leaders—all who made decisions about the canal—met to coordinate.

"We'd meet together for about an hour, usually about 6:00 A.M.," Terry said. "Then Tom Tate would be off and running. I'd stay at the office and handle the press."

Oakley reservoir had become national news. "First it was the *South Idaho Press*, then the Twin Falls paper, then TV, UPS, AP—people I'd never heard of before," Terry said. "Tom never stayed around. He didn't want to take time talking to people."

Kent Foster avoided reporters. "My job was technical," he said. "I didn't want to be interviewed. Terry handled the press and did a great job of that. Someone had to do it."

A reporter from the McNeil-Lehrer NewsHour found Marvin Blacker at the court house and asked him if they could film the digging of the canal through his farm. "I told him no," Marvin said. "I came home and they were already there with their cameras."

Police Chief Leman Messley furnished officers to accompany the visiting VIPs. It was election year.

"We had some legislators who were roaming around," said Kim Cranney. "Nobody wanted to talk to them. One guy wearing a suit came by where I was working. He said he'd do anything he could to help us. I told him if he'd get out of that suit he could help sandbag."

When people called and asked Rex Gerratt to come to the reservoir to have his picture taken with a VIP, "if I was headed for the reservoir, I'd turn around and go the other way," Rex said.

"There were lots of people from the media," said Earl Corless; "CBS made a video from a helicopter. A Boise TV station interviewed me on the site. They wanted Straubhaar, couldn't find him, decided to talk to me. These media people get their story, and then try to fit you into it. A reporter tried to get me to say the dam was in danger. She kept asking me and asking me about seepage and if the dam was going to break. I told her finally she was asking the wrong question—the real problem was all that water running uncontrolled over the spillway. Later I found that a photographer had gone to the spillway and taken a picture of little seeps of water coming out. They wanted to use the picture as background, were trying to build their story around it.

"What most people do not understand is that a dam is not an impenetrable barrier," Earl said. "Water flows through all dams. The dam is designed to take it. If seepage increases dramatically, if it begins to bring material out, then it's cause for alarm."

Later Earl called Burke Scholer on the radio to tell him about a meeting. "Before I could tell my supervisor, KMTV Channel 11 called to see if they could come," Earl said. "We knew then that everyone was listening to our radio conversation. After that, instead of saying 'leaks' or 'seepage' we talked about 'cipolletti weirs'. A cipolletti weir is a device for measuring flow. We'd say, 'cipolletti weir number two flowing at so much.' That way anybody who did not know water would not get the wrong idea."

Kloepfer trucks and farm trucks hauled riprap (rocks and other fill material) to reinforce the canal banks. "We hauled rocks to near the bean warehouses," said Orvin Jenks. "I drove one of Adams' trucks. We weren't sure they'd get the canal done in time. We were nerved up about it."

A lot of farmers called, "extolling the virtues of their field stone," remembered Terry Bingham. "Kent Searle had a mobile phone. People would call up and say, 'I have this real good stone here, high quality, easy access.'"

"From the time this county was taken out of sagebrush, rocks plagued the farmers," Rex Gerratt explained. "In the early days

Loading field stone to line the canal banks. (Photo courtesy of Leroy Jarolimek.)

they hauled rocks to the fence line with horses. The best place to take rocks was to the desert ground, to the brush ground, and pile them up. Hundreds of thousands of tons of rocks were hauled off farm ground. For a hundred years people had been cursing those rocks, thinking they had no value. They were all stacked out in the desert—some of the piles were six or seven feet high. Those rocks were a lifesaver. Loaders were dumping rocks in those trucks all night long. There was a lot of laughing about that—except for the poor guy who was getting the rocks on his land. He wasn't laughing."

For several years Fred Hawker had piled rocks in a corner of his field. Early in the spring he had paid Kloepfer's to haul the rocks to the desert. When they began building the new canal, "they hauled in rock and put it in the exact same place where I had all that other rock hauled away."

Volunteers piled rocks to reinforce some roads in the natural drain. The road would act as a buffer to slow the water and would give people a place to cross. Dam safety manager Dave Hollingshead went with Terry to look at the roads. "I liked to kid my LDS friends about making water run uphill," Dave said. "I remember looking at one road. I said, 'You'd better be of the right faith, brother, if you think this one is going to work.'"

South of the natural drain all roads were cut. "Most of the roads are being cut as they go because there won't be time to return," wrote Donald Clark. "Some of the homes are being isolated in the Oakley area tonight."

Lloyd Brothers Construction erected concrete bridge piers in the canal near Oakley's main street. They started on the piers as soon as they finished the spillway, Max remembered. "We had a deadline—worked long hours to get it in." They planned to cover the piers with planking after the water started flowing.

The highway districts posted road-closed signs and erected barricades. "I had one guy—all he did all day long was make road-closed signs," said Oakley manager Ray Kimber.

"We put up all the road-closed signs we had," said Burley manager Neal Thomas. "We had Leonard King make twenty-six big road-closed barricades."

"We had the county cut in half," said equipment operator Lawrence Burch. "There were a lot of mad people at certain times. I was digging through the main road going west of Oakley. A car got right in my way—just parked there. A lady jumped out—it was the post office lady, delivering mail. She was mad because I was tearing up the road. She wanted my name and everything so she could report me. She said, 'How am I going to get across?'

"I said, 'I guess you'll have to go to Highway 30.'"

"After we was cut off, I had to drive sixty miles to get to Burley," said Ed Clark, who worked for Cranney farms. "I'd have to go clear to the dam, cross the bridge there by the gates, and then come back. I was the gopher. I'd go fer anything that was needed. I hauled lunch to the guys, and fuel to them on both canals."

"The kids enjoyed it," his wife Thelma said. "They couldn't go to school. But they worked—the little ones and the big ones. There was a bunch of kids at the gravel pits and a bunch here on the corner where sand had been piled up. It was kids and shovels. They filled sandbags as fast as they could work. The older ones drove tractors with trailers taking sandbags to all the homes.

"I can't remember how many trips Ed made to town to get shovels—more shovels and more shovels. He got all there was up to Burley. Afterwards not a shovel came back to Cranneys. The kids was so sick of shoveling they didn't want to look at another shovel ever again."

Oakley Stake counselor Dallan Elquist supervised the sandbagging. "You've never seen kids work so hard in your life," he said. "I didn't have to reprimand anyone for monkeying. When they let school out, the kids knew it was serious."

Norma Elquist helped some young girls fill sandbags. "They would have one or two holding a sack and one shoveling," she said. "They'd trade off after so many bags. They didn't have enough shovels, so two girls and I scooped it in with our hands. Everyone was excited to help."

Wayne Mullen and Luther Gardner took a load of teenagers and sandbags to Mitchells' ranch on Goose Creek above the reservoir. "They helped us make a good wall to divert the water away from the bank so it wouldn't wash out more," wrote Ellen Mitchell. "We surely appreciated their help."

Fourteen-year-old Shay Watterson ran a scraper for his father Lynn. Other teen-agers drove tractors to pack the banks. "They didn't want to get off even to get something to eat," said Doug Adams.

Delma Bean was not concerned about her fourteen-year-old grandson driving a tractor. "They'd worked on tractors in the fields, knew how to run them," she said.

"It was not dangerous," said Larry Gerratt. "Everybody was dualed up—had two sets of big tires on the back for stability."

Stake Relief Society President Suzanne Critchfield made a list of families in the danger zone. Ward Relief Society presidents arranged for emergency housing. "Everything always seemed under control to me," said Suzanne. "There was no panic. President Gerratt was always reassuring, always said we'd make it. If that feeling comes from the head, it goes clear down the line."

"We were asked to draw an evacuation map for our ward," said Randy Hardy, counselor in the Oakley Second Ward bishopric. "Six to eight men were assigned to call various families. They had to be in personal reach of them in case the telephones would go. The canal split our ward in half. We had it divided so each side could take care of itself. "View Second Ward was assigned to be the assistant for our ward alone. If we needed loaders, trucks,

Shane Beck, a high school junior, driving a tractor to pack the canal banks. Eight-year-old Kody Beck is looking out the back window. (Photo courtesy of Susan Beck.)

people to do things for us, we called on them. They had homes arranged in View Second in case we had to evacuate. Those providing housing knew exactly which family would be coming, how many kids they had, how many beds they'd need."

Some people in the danger zone moved furniture and cleared basements; others did not think it was necessary. Leon Adams, returning from his fishing trip, was surprised to see people with their pickups piled high with their possessions.

Construction crews dug along one side of the railroad, crossed the tracks at Randy Hardy's place, and dug along the other side. "We went down the track with torches," said Wayne Mullen. "First we cut the rails every other length so Kloepfer's loader could lift them aside, but we couldn't do it fast enough. Then we cut them in hundred-foot lengths and the cats pushed them out of the way."

Leon Adams guarded his main irrigation line as the equipment came through. "I stood there on that mainline until they got by it, so that I could protect it," he said. "The last cat nicked it, knocked a hole in it."

At Kenyon the route left the railroad tracks and cut across farms. "That second day as we was going through people's fields,

Digging through Fred Hawker's newly planted field. (Photo courtesy of Fred Hawker.)

you could see people looking at you," said Lawrence Burch. "I had a feeling they thought it was a dream—not reality."

"Bruce, Dad, and I were farming together," said Roland Bean. "If the water had come down Goose Creek, it would have come through one farm. The new route came through 2¼ miles of our Kenyon farm. It came through a mile of sugar beets—they were up and looking so pretty—a mile of potatoes, and a quarter mile of wheat."

Val Robins was dismayed to see the size of the canal. He had expected it to be about twenty feet wide, the size of the third-lift canal.

"I thought it would be like a ditch, three or four feet wide," said Gloria Robins. "When they started digging it, it was ninety feet wide at our place. I started to cry. It wasn't a ditch, it was a river going through our farm. I had no idea it would be that size. I remember somebody hugging me. I bawled and bawled."

"That was the tragedy of the whole thing," said Larry Gerratt. "People who had to watch it come through their places, see grain pushed out. It made me feel so sorry for them. I could relate to what they were feeling inside."

"It was the prettiest field of grain," said Fred Hawker. "It was green wall to wall before they went through."

J.J. McLaws was surveying when a farmer came running out and said, "You're not going through here."

"The guys on cats looked at me," said J.J. "They said, 'What are we going to do?'

"I said, 'I guess we stop here.'

"Rex Gerratt came by. He took the farmer in his pickup, and they drove off about a half a mile. In a few minutes they came back. The farmer said, 'Okay, let them go.'

"I never seen anything like it," said J.J.

Rex remembered the incident. "At one point it looked like there were two ways the route could go," he said. "If we went one way, it would damage one farmer; if we went the other way, it would hurt the other. The farmers were both yelling, 'Don't take it through my field!'

"By the time I arrived each had his own little group gathered around him, ready to take his part. I had the two farmers sit in the van with me. The windows were rolled up so no one could hear us and we were away from the crowd. I explained to them that we had to go the way that would help the majority. We had to be fair. After a few minutes they calmed down. The affected farmer was willing to let it go where it had to go."

"May 18," wrote Julie Randall in her journal. "By 10 A.M. the crew was going strong. They hauled the dirt around to dam up the silage pit. They brought in sandbags and put around the well, our cellar, and our houses. I took crispy treats and Kool-Aid out for them. Bruce Beck came and helped us plant our corn. Somebody else was plowing. There were lots of rumors. We thought we were going to be flooded out. I packed a suitcase—had stuff in the car ready to go."

The Harrises moved everything out of their basement. "We had it all stacked in the living room," Marian said. "It was hard to get around. The two stakes came running out to sandbag—sandbagged the wrong side of Jolene's house. My father went to live with my sister Barbara Graham. It was too traumatic for him."

Near the diversion gates below the dam, Curtis Holmes drove his D-7 cat to clear the brush. "Kloepfer foreman Dave Peterson told me to do it," said Curtis. "There were cement ditches and all kinds of stuff. He said, 'It doesn't matter, drive over them. They'll all get washed out anyway.' I knew he was right, but I had the hardest time doing it. It went against all my morals to drive over those ditches and cause all that destruction."

At the opposite end of the canal near the Snake River, the natural drain crossed under Highway 30 and the Union Pacific Railroad. "They had culverts under the railroad already," said Kent Foster. "The railroad people told us what size their culverts were, but they were bigger on either end, restricted in the middle." He knew they would not be large enough.

"I was assigned to handle the railroad," said Police Chief Leman Messley. "I called the Union Pacific offices at Pocatello and told them the problem. I didn't get an appropriate response. I called again, told them it was a serious situation."

The next day two railroad officials arrived. "I gave them a car and they took a tour. They came back, gave me the keys, and said, 'We'll talk to our supervisor.'"

Leman called the supervisor who told him that the officials did not think anything needed to be done. The culverts appeared adequate; the emergency was not that bad.

Leman then called someone at Union Pacific headquarters in Omaha. "I asked the guy if he'd heard about the problem," Leman remembered. "Well, he'd heard a little about it, he said.

"I said, 'This thing is progressing and we've got to have action.'

"'We'll get around to it,' he said.

"'It's got to be done before Sunday,' I told him.

"He said something about not wanting to be bullied.

"I told him, 'I know farmers who know how to use dynamite.'

"'Are you threatening—' he started to say.

"I said, 'I'm just telling you that you'll move that railroad, or I will.'"

The police powers declaration gave Leman the right to damage property if necessary.

"I'd have done it too," said Leman. "We could not allow the railroad to jeopardize our entire project."

Within hours Pocatello dispatched a big bulldozer and began moving track.

While the civic leaders and the Corps of Engineers supervised construction, LDS church leaders organized the volunteers. Be-

Installing culverts under the Union Pacific Railroad and Highway 30. (Photo courtesy of Phil Hodge.)

cause he was bi-regional emergency director, Oakley Stake President Rex Gerratt had authority when emergency situations involved more than one stake. "It worked out real well," he said. "Here I was, the emergency director, and the emergency was in my stake." On Wednesday night he called a meeting for the stake presidencies to explain about the canal and ask for their help.

"It seemed like such a monumental task," said Paul Stake President John Stevenson. "I remember thinking, 'There's no way they can get a canal built that fast.' But everybody was so willing. Rupert Stake President Jack Duncan and I told President Gerratt we'd help in any way we could."

They needed loaders, rippers, tractors, and dozers. "President Gerratt met with us at five in the morning," said President Stevenson. "Those brethren in charge—it seemed impossible that they had the strength to keep going, to do all that they had to do. At 4:00 or 5:00 A.M. we'd be getting calls and making calls."

Burley, Burley West, and Oakley stake presidencies and emergency directors met each night, usually from 10:30 to midnight. One night they met at 10:00 P.M. "I don't see why you want to meet so early, and break up the whole afternoon," said counselor Bob Sagers.

Oakley Stake historical clerk Donald Clark took minutes of the meetings. "President Gerratt asked me to make a record,"

Donald later said. "He came to me and said, 'We need to have a record kept on a daily basis. We're making history here now. I want you to keep track as near as you can of what's going on.'

"We'd all sit around the table in the Burley Stake Center. The emergency directors would come in and report. Each had assignments. They'd take it item by item, bring up the problems. They'd talk about what happened today, what needed to happen tomorrow, how many people were working, how many were needed the next day. This is what I tried to write down."

"We'd be so tired when we'd get to the meetings," said emergency director Wayne Mullen, "but there was no time to waste. There was such a feeling of urgency."

"It was reported considerable accomplishment today on the canal from Oakley," wrote Donald Clark on Friday, May 18. "By dark it should be nearly complete for 11 miles. . . . The Corp of Engineers will be working all night. Roads in this area will be closed by morning. . . .

"President Petersen reported inadequate markings and knowledge of the channel in his area. He requested help in the morning to better define any detail work that must be done. . . . He also reported on making contacts with land owners in the lower channel area and finding mostly good feelings. . . .

"Areas of Concern: The flat area on the Robins and Blauer Farms and the soil type may make packing of the canal banks difficult but highly essential. . . . Heavy packing tractors with duals are still needed to assist in packing the complete channel.

"Roads in the natural channel area will be left at the discretion of Terry Bingham. When left they should be riprapped for protection of the road. It is hoped the water may be slowed somewhat. A need for about 100 men to riprap on standby was expressed by President Gerratt, 30 from each of the Burley stakes and 40 from Oakley. . . .

"Assignments recommended by Pres. Gerratt for tomorrow are: He will be over the whole working area. Pres. Sagers will be responsible for Murtaugh. Pres. Elquist is in charge of the Oakley youth and sandbagging. Burley West Stake from Highway 30 to Randalls. Burley Stake from Randalls to Beans. Brother Mullen sandbagging at Merrill Warr home, and distribution and control of 20,000 sandbags in Oakley. Need 2 days' lunches for Oakley Stake or 300 made tomorrow, Burley West 100, Burley 100. President Sagers will handle an afternoon pop run."

Although the stake presidencies discussed what needed to be done, as religious leaders responsible for the people within their stakes, they were most concerned about receiving guidance from the Lord. They opened and closed their meetings with prayer.

"We expressed gratitude in our prayers," said President Petersen. "We were grateful for people's unselfishness and willingness to do what had to be done. We prayed for the weather. We prayed about the decisions, that those who had to make decisions would be inspired, that they would do the thing that would be of benefit to the most people and cause the least damage to the fewest."

"We were calling on the Lord and needing His help," said President Gerratt. "Our prayers were not just ordinary prayers. We were on our knees pleading for direction about the problems of the day. The direction always came in time. We were being touched to know what the best judgment was."

"President Gerratt said, 'If you see me making wrong decisions, tell me,'" remembered counselor Dallan Elquist. "We didn't have to. We all felt that President Gerratt was inspired."

Wayne Mullen remembered how impressed he was with the stake presidents' concern for the people. "One person could not be so honored to meet with all those inspired brethren," he wrote in his journal. "I developed a great love for each."

Saturday at 9:10 A.M. the water in the reservoir reached the level of the old spillway. Without the spillway addition, it would have been pouring over the top.

The U.S. Geological Survey measured the inflow from Goose and Trapper creeks; the Bureau of Reclamation measured the increase in the reservoir, and the Department of Water Resources measured the outflows. Only Bureau of Reclamation officials worked full time at Oakley. The others also monitored flows at Raft River, Salmon Falls, Dry Creek, and the Snake River.

The USGS officials had problems measuring the flooding creeks, remembered Loren Holmes. "One guy at Salmon Falls was making a measurement, when here came a whole mass of rattlesnakes all rolled together in a ball. A den had washed out."

Earl Corless spent most of his daylight hours at the reservoir. "I'd take gauge height readings at fifteen-minute intervals, hourly readings off Goose and Trapper Creek," he said. "I walked the face of the dam and the spillway. I was watching for seepage, for muddy water, for any change in movement. Allen Merritt from the Department of Water Resources measured the seepage with a current meter to establish a data base. There was major seepage, but not more than what we expected."

Every two hours Earl called Denny Davis to give him the figures. "I keyed them in our computer," said Denny. "That gave information to everyone who had access to our computer. I plotted readings and hand-carried them to the sheriff's office to Terry Bingham. The press called, and I gave them the numbers. I even got a call from the *Philadelphia Enquirer*."

"Earl and Denny deserve a pat on the back," said Burke Scholer. "They kept us abreast of the water situation. The water had to be turned out on a given date based on outflows and inflows. They had it worked out exactly. Earl said the water would peak at the middle of the second board on the spillway."

As Earl walked the reservoir Saturday morning, he noticed some driftwood floating near the surface. "It did not look like normal driftwood," he said. "Part of it was lower than the water surface. Then I noticed a vortex—a little whirlpool. Water was going some place, and I didn't know where. It might not be anything, but it could be the first sign the dam was failing. I called Terry and Ernie and my boss Don Tracy to come and look at it."

They stood there, watching the driftwood go round and round. It had not changed since Earl had first observed it.

"Ernie said, 'Do you want me to bring a truck up here and cover it with dirt?'" Earl remembered. "I told him no. If it was bad, a little dirt wouldn't help. Besides if we covered it up, we'd never be able to tell where it was going.

"Terry said, 'Do I evacuate the county?'

"I looked at my boss, and he looked back at me, and I knew it was up to me to make the decision," said Earl, "as hairy a decision as I've ever made in my life."

If the dam was not in danger, moving people would cause panic and hardship and would jeopardize their finishing the canal.

Yet everyone remembered Teton. If the Oakley dam were to fail, they had to get people out of the way.

"I had inspected everything I could think of before that little meeting," Earl said. "Everything else was normal. I had that in my mind as a data base."

Earl had years of experience with water, knew what it would do, knew what the signs would be if a structure were starting to fail.

"I told them not to evacuate," he said. "I told them I didn't think it was that big a problem.

"As soon as I said that, they scattered like a covey of quail. I remember standing there all alone, looking at that vortex and thinking, 'I sure hope I made the right decision.'"

The work continued. Since Thursday morning Ernie Straubhaar had been at the site constantly. "I was there twenty-four hours a day," he said. "I didn't dare leave—there were too many problems. When I'd get too tired I'd roll over in my pickup and take a nap."

Terry Bingham was worried; they were behind schedule. Dam safety manager Dave Hollingshead told Terry they had to release the water by 5:00 A.M. Sunday whether or not the canal was completed. Farmers with rippers worked in front of the equipment, loosening and ripping the dirt so cats could move it faster.

A McNeil-Lehrer camera crew took pictures of Marvin and Lovel Blacker waiting for equipment to arrive at their farm. If the canal was not finished on time, Marvin feared that all the water would be dumped on them. Corps officials told him the cats would be to their place Friday afternoon. Saturday at 3:15 P.M. they were still waiting. Marvin showed the camera crew the stakes across his field where the canal would be. "There's a mass of water that's going to be coming," he said. "How we're going to handle it, we don't know. I guess in the end it will all come out, we hope. I hope I'm big enough to make it through." He tried to speak calmly, but the camera caught his worried face.

Part of Ernie's crew moved their equipment from the almost finished canal near Oakley to the natural drain. "They came with big machines and ruined our holler," wrote Julie Randall. "It was such a nice, peaceful spot, fun for picnics. They destroyed it in a few minutes. It was a bad day at our house."

"They tore up the asphalt roadway, tore up the trees, tore the front of the cellar off," said Richard Randall. "It was hardest on Dad, to see them pulling things out and ripping things up, knowing that he'd probably never be able to build it back in his lifetime. They took out trees that had been here since the place was first homesteaded. They took out seven big trees and a bunch of little ones."

They went through Jack Gochnour's place, uprooting trees. "It was very traumatic," Jack said.

"We didn't really understand what it would be like until we saw the machines," said Marian Harris. "When they moved to our property, it was kind of a shock."

"I stood out there and watched them go through our farm," said Willes Blauer. "It was late in the afternoon. People think it's just dirt, that's not the case."

Al and Kathy Barrus moved everything from their basement and sandbagged their house. Kathy was not frightened. "We were so busy working," she said. "I didn't stop to think about it—didn't know enough to be scared. We focused on what had to be done. We had faith because President Garratt was working so hard. We knew he was working under inspiration."

The water would be two hundred yards from their home. "At first they did not want to cut the 200 West road, wanted to leave it to act as a buffer," said Al Barrus. "I was afraid it would flood my house. You get a real protectionist feeling when it concerns your home and family. I cornered the guy from the Corps who was calling the shots, asked him if they were going to break up the road. He said no, thought it would be all right. I said, 'If that were your house sitting there, would you break up the road or let it go?'

"He said, 'We'll break up the road.'"

Canal company secretary Klyda Wake moved from her house. "I had a big canal on either side of me," said Klyda. "When a friend asked me to spend the night at her place, I was glad to do it."

Until Saturday the Van Tassell family had not worried about the canal to Murtaugh which ran along the side of their field. "We thought we were fairly safe," said Carl Van Tassell. "We thought they'd fixed it to take care of us."

Saturday afternoon about four o'clock Kent Van Tassell went to look at the canal. "He was just curious," his mother Elda later remembered. "He was not concerned about it." He saw the water begin to break through the canal bank and run into their field.

"We tried to stop it with straw bales," said Carl. "It kept getting worse. We could see that we were losing it. We had a number you called on the telephone and they'd get hold of Leland. The Bakers came out with their equipment. They tried to stop it with earth first. That didn't do it."

That evening the two Burley stakes were holding conference. It had been one year since Burley Stake had been divided and Burley West Stake organized. Elder Hugh Pinnock, a member of the Quorum of the Seventy, had come from church headquarters in Salt Lake City to preside over the meetings.

Someone called President Petersen to the telephone. Because he was a medical doctor, his congregation was accustomed to seeing him leave meetings. This time he returned and announced, "That was an emergency call, but I'm not leaving, some of you are. The canal to Murtaugh is breaking out and they need men with dump trucks to haul riprap. Bishop Harry Shaw in the north lobby will give you directions on how to get there."

Several men left. People in the meeting could hear a low buzz of voices in the lobby; then the men were gone.

"Within thirty minutes about 150 people were out here," said Carl Van Tassell. Ernie Straubhaar sent Kloepfer trucks and a loader; farm trucks from Oakley Stake came.

"A lot of people brought other people's trucks," Carl said. "LaMar Anderson had someone else's truck. One woman had her two little children in their night clothes with her. She was in a borrowed truck hauling rock."

"It beat up those trucks pretty good," said David Beck. "I had a dozer. They'd dump the rocks and I'd push them in. They were trying to keep the canal intact. It would break out that quick."

"The banks weren't packed, they were just pushed up," said Susan Beck. "It was like the water melted the banks away."

The water had run into part of Van Tassell's field, making a muddy area for the trucks to cross. "I spent all night dragging trucks through that muddy spot with my four-wheel-drive tractor," said Carl. "Finally the mud got so deep in the field they had them come in from another way."

Kent Van Tassell and Rulon Van Tassell hauling straw bales to plug the leak in the canal. (Photo courtesy of Elda Van Tassell.)

The men from Kloepfers worked until about 11:00 P.M., Carl remembered. "Then they said, 'We're give out,' and left. Their trucks were parked here. My son Rulon found a semi with the keys in it and drove it. Duane Child drove their loader.

"The next morning Kloepfer's main man came out here, asked who drove that truck. 'I did,' said Rulon.

"'Ever drove one of these before?'

"'No,' said Rulon.

"'You mean you drove down a canal bank at night and hauled rock and you'd never drove one before?'

"'Yes,' he said, by that time wondering if he was going to be hauled off and thrown in jail.

"The man said, 'Do you want to come and work for Kloepfer's?'"

City of Burley trucks hauled rock, remembered Olene Spurgeon. "They were big rocks—it was all our loader could do to lift them. One person got lost. He was driving around with a load of rock—didn't know where to dump it."

Trucks from the Burley Highway District hauled rock. "They wanted to run our trucks more," manager Neal Thomas remembered. "They asked me if they could have some other people drive. I said okay if they could find experienced drivers. I didn't want them pulling just anybody off the street. They got people we knew, super good drivers. We worked twelve-hour shifts, with half my guys on each shift. Containing Cottonwood was really a fight."

On the new canal near the Snake River, railroad and state highway crews removed the track and dug through Highway 30 to install two new four-foot culverts near an old concrete culvert. They had placed the new culverts and were preparing to cover them when Rex Gerratt and Gordon Luke arrived. Rex and Gordon watched the work a few minutes, then turned to leave. It was dusk.

"As we left there, I glanced in my rearview mirror," Rex said. "Through the mirror I sighted the windows of Lane Gochnour's dairy barn lined up with the grade of the new culverts being installed. The mirror, in fact, acted as a transit. I knew that with the culvert that high, it would flood the dairy. I turned right around and Gordon and I climbed up on top of the culverts."

The highway and railroad engineers were directing construction. Rex asked them to lower the culverts. "This man's dairy will be flooded before the water goes through these culverts," he said.

"There were a lot of adjectives spent in answer to my request," Rex later remembered. "They wanted to know who I was and what authority I had to question what they were doing. They told me in no uncertain terms that the culvert was going to stay right where it was at, and besides that, they could not lower it because it was sitting right on top of some natural gas lines.

"Ironically, the day before I had checked the gas lines in relation to the old culvert. Because of that, I knew the new culverts could be lowered three feet which would save the dairy. The highway engineer said, 'You don't know what you're talking about. We are not taking these culverts out and lowering them.'"

They argued a little more. "Finally the railroad engineer told the highway engineer that his transit was already set—to quit arguing and look in it," Rex remembered. "When he looked, it was very obvious that they were going to flood the dairy.

"Then he said, 'It's only a barn.' But it was that guy's livelihood. I told them we had not deliberately flooded anybody so far. We were not going to do it now."

Someone called the scs engineer, who checked the culvert level. "The engineer used the phone in our barn," Lane Gochnour said. "We could hear him talking. We heard him say that if they left the culvert where it was, it would flood to the middle of the barn windows. It was scary to hear what they were saying. It would flood the barn, might flood some houses, could damage Ore-Ida."

Culverts under the highway and railroad. (Photo courtesy of the Walla Walla District, U.S. Army Corps of Engineers.)

It was time for the stake presidencies meeting. "Rex went to the meeting—left me there to see that they took care of it," Gordon Luke said.

The Burley and Burley West stake leaders, who had finished their evening conference, were dressed in suits. Elder Pinnock from Salt Lake City was there. "I went to the meeting in my old muddy clothes," Rex said.

"President Gerratt reported having many ups and downs today," Donald Clark wrote. "The biggest problem is keeping the water in the Oakley-Murtaugh Canal. The banks are washing extremely, many loads of rock are being hauled. . . .

"We are near completion of the Oakley-Burley Snake River Canal. . . . The present plan is to turn out 20,000 [miners] inches of water from the Reservoir Flood Gates at 5:00 A.M., Sunday, starting with 5,000 inches and building throughout the day. . . .

"We need to set up a canal bank watch, to be effective as soon as the water gets in the channel. The experience of the West Canal tells us it is essential; however, its banks were not packed, this should make a difference. . . .

"The two additional culverts being placed in the Highway 30 and Railroad Crossing . . . are a big concern. . . . It was felt that even though we're tired, a delegation should check to see the progress. . . .

"The 3 Stake Presidents, with Elder Pinnock and about 5 others from the committee, viewed the site to find the culverts being placed without being lowered," Donald Clark wrote. "President Gerratt immediately called Alfred Barrus, our County Prosecuting Attorney; the group went to his office and contacted the Superintendent of Highway construction."

"We had a meeting at my office, went out to the site, came back to the office again," said Al Barrus. "We had a real headache trying to get to the right people."

"We got hold of the man in charge," remembered Dallan Elquist. "He hemmed and hawed, said he'd have to get hold of somebody to check it. We said, 'We haven't got time for someone to come and check it.' He didn't want to lower it—said it was against regulations. I don't remember what the problem was exactly—something about the amount of backfill on top of a culvert.

"President Gerratt said, 'I don't give a darn about regulations, put that culvert where it belongs.' The superintendent finally made the call to the people putting it in."

Al Barrus went home while the others returned to the site.

"They had lowered it a little, but not enough," said Gordon Luke.

"I barely got home when they called me again," Al said. "We went back to my office."

"We got the state highway guy out of bed again and he was mad," said Rex. "I told him, 'If you bury those culverts the way they are, we'll plow them up as fast as you bury them.' I told him I was not trying to be ornery, but we hadn't flooded anybody yet, and we were not going to start now."

"We finally went over his head, called Boise," Al remembered.

They went back to the site. "President Gerratt would not leave until it was changed," Al said. "He was adamant. He was more than adamant. He said, 'You can either do it right, or it's not going to be done, because we won't let it happen.' I think he felt such a weight because he had pushed for the canal to go that way. It was the right way to go, but it would have been tragic if the water had backed up there for a couple of miles."

"It was 2:00 A.M. when the workers agreed to lowering the culverts," wrote Donald Clark, "so we all went home, except President Gerratt and Gordon Luke who made a trip to Golden Valley to check the West Canal."

"It was Rex Gerratt's stubbornness that got it done," said Gordon Luke.

"He exhausted them," said Lane Gochnour. "He stayed there until three in the morning. They finally did it out of exhaustion."

Rex dropped Gordon off at his house, then returned to the highway for one last look. He walked over to the construction area. The culverts were being lowered, not as much as he had wanted, but enough that the dairy would not be flooded.

"As I started to leave, I turned and stepped down a little grade," Rex said. "You know how you do when you think there's another step and there isn't. That threw my knee out—I've got bad knees anyway. They helped me to the top."

He did not go to a doctor; there was no time. "I just hopped around with it," he said. "I did not slow down."

"But he was in pain all the time," his wife Marjorie remembered.

Water entering the new canal.
(Photo courtesy of Nona Jenks.)

Chapter Eleven

Sunday, May 20

Early Sunday morning Larry Gerratt and Bill Garrard went to the dam. They needed to arrange a schedule for the volunteers who would patrol the canal. The water at the reservoir was about six inches high on the spillway addition. "Off to the side we could see the water flowing to Murtaugh Lake," Larry said. "In front of us was the canal to the Snake River. It hit me for the first time—the vastness of what had been done. 'We did it,' I remember thinking. 'We built those canals and saved our valley.'"

Though crews and equipment were still working on the lower end of the channel, the officials planned to release the water at 5:00 A.M. "I told them, 'We can't sit on this any longer,'" said dam safety manager Dave Hollingshead. "We were running out of board and out of time." Dave estimated that it would take about twenty-four hours for the water to reach the Snake River. The work could be completed before the water reached the end of the canal.

Two brass gates in a natural rock wall near the bottom of the reservoir controlled the flow of water into a tunnel. Two steel stems connected the gates to a gear system 150 feet above the gates which raised and lowered the gates and stems. The stems were hooked to the side of the rock wall every twenty feet with concrete.

Occasionally a rock under a gate would keep the gate from shutting and cause part of a stem to pop out of the old, crumbling concrete. Through the years the stems had been repaired several times, the last time in 1981. "It was a blessing that the stems had been repaired that recently," Rex Gerratt said. "The gates might not have held without those repairs."

Even with the repairs, the officials worried about the gates as they prepared to raise them higher than they had ever been raised before.

"The reservoir was backed up eight miles, 135 feet high," said Winslow Whiteley. "There was tremendous pressure at the bottom of the dam. We were so worried about washing the gates out. If the gates had gone, there would have been no controlling the water."

"We were concerned about opening them up and getting them jammed and not being able to get them shut again," said Allen Merritt. "It was always a concern when they wanted to release more water."

"We opened one gate a little, then opened the other," said Dave Hollingshead. "We had to do it by trial and error."

The water ran through the gate opening into a tunnel under the floor of the reservoir and then into a canal. Diversion gates a mile below the dam divided the water into the East and West canals. A third set of gates, which had not been used since 1921, would allow water to flow down Goose Creek channel to the new canal. Years earlier those gates had been cemented shut. "People were there with jackhammers trying to break up the cement," remembered Loreen Gorringe.

"When the order was given to let the water go, they could not turn the gates," Terry Bingham said. "They were all rickety, groaned and creaked. We were late running the water because the gates would not turn."

Sunday at 8:04 A.M., water began flowing down Goose Creek channel. They released 100 cfs initially, gradually increasing the flow to 400 cfs. Dave Hollingshead directed the releases. "It was mostly seat-of-the-pants determination," he said. "I was basing releases on historical accumulations of the snow pack, on approximations of volume and flow rate. I knew how much capacity remained in the reservoir. We wanted to equalize inflow and outflow. It was a real balancing act to release a sufficient amount and not exceed the capacity of the canal to handle it. The SCS kept saying, 'It's designed for 500 cfs,' but we later released almost twice

Jay Gorringe and Teach Wilson trying to open the old Goose Creek diversion gates. (Photo courtesy of Ryan Thomas.)

that much. I was down below the dam as Jay increased the flow. We'd open it, wait to see how many people were screaming down the way. They weren't screaming too bad so we'd open it some more. There was real concern about handling the water."

"When they turned the water loose, they took me and Glen Straubhaar, Ernie's boy, up to the head of the canal," said Lawrence Burch. "We went down the canal on loaders, checking for debris, cutting high spots and filling low spots, making sure there was not an open culvert leading to an irrigation ditch. Sometimes the water would get a little ahead of us. The base was gravelly enough that we could work in a little water without getting stuck." Several people remembered seeing them in the canal, working in the water.

Fritz McCoy and Lynn Watterson were still working on the canal banks, remembered Lynn's wife Paula. "Lynn was working on the east side and McCoy was on the west because he was from Twin Falls. Lynn hollered, 'Get back on your side, here comes the water!'"

The water flowed for one mile in the old Goose Creek channel, past Merrill Warr's home, and then entered the new canal.

Watching the water enter the new canal at the Merrill Warr place. (Photo courtesy of Steven Pearson.)

"Sunday, 20," wrote Nona Jenks in her journal. "They closed church. I got up at 6:15 because they said they would release the water at daylight."

Nona had her camera ready. "Just waiting! waiting! waiting!" she wrote.

At 8:30 A.M. at Golden Valley road, she watched crews pouring concrete abutments for a bridge. "Denver Altom said, 'Hey, guys, we better hurry up. Mrs. Jenks says they've let the water go.'"

"At 11:30 A.M. I went back to Oakley," she wrote. "A guy on a moped said the water was still quite a ways away. I decided to walk down the canal bank and look for it. I walked to the Merrill Warr home, saw all these people. A guy from KUTV in a helicopter landed right by us. One of the news guys talked to us ladies. I watched the water come through the pasture of Kelly Poulton's and across the road to the pasture on the Merrill Warr place. The water reached the man-made channel at 12:45. It isn't coming too fast yet. I could walk faster than it was moving."

"It was like a race," said Bill Crystal. "They had spotters all along the way telling people, 'Well, it's here. The water just got here.' I think the people who were not farmers were surprised that the water took so long to come."

"There was great excitement on Sunday," said Dave Hollingshead. "Some people were in the channel on three-wheelers, moving ahead of the water. Other people would find a road to the channel, sit and wait for the water to come, wager when it would come through."

"It is very scary," wrote Marjorie Gerratt. "They let the water out of the bottom of the dam midmorning and it slowly started to Burley. It doesn't look slow—it's really moving."

All the major news networks sent reporters and photographers. Only people who lived in the area or those on official business could cross the police barricades on the roads. "There were lots of sightseers," said Reserve Deputy Doug Adams. "Most of them were just curious. When I told them they couldn't go through, most would visit for awhile, then they'd turn around and go back. The public was good—at least the ones I dealt with."

"They were stopping people every mile—especially along the Oakley highway they were stopping them," wrote Nona. "No one from out-of-town could come. My son from Blackfoot came and couldn't get in. Actually there was no way to get through, no way to cross. I talked to people and walked around. It was windy and cold."

Earl Corless watched the reservoir. The vortex had not changed. The Boise office had sent dye which he and Don Tracy poured into the vortex. "We put the dye in early Sunday," said Earl. "It took a long time for it to get to the stream. It was going through a conduit through some rocks—the same place that it had been leaking for years. Then we knew we had no problem."

Meanwhile volunteers who had worked all night on the Murtaugh channel were still riprapping. Carl Van Tassell did not know they had turned water into the new canal. "There was no church," wrote Carl in his journal. "Everybody worked on the channel. I helped unload the rock and they went up six or eight feet on both sides."

"They got it stopped on our side, and then it started breaking out on the other side, onto Wolvertons," Carl said. "They'd get it stopped at one spot and it would break out some place else. By Sunday noon they had riprapped both sides of the canal for two miles, from here to the Golden Valley road."

More church groups arrived to help. "A bunch from the Unity area brought all their youth out," David Beck remembered.

When the men arrived for Priesthood meeting at the View Second Ward, they were sent home to change to their work clothes

Riprapping the Murtaugh diversion canal. (Photo courtesy of the Soil Conservation Service.)

so they could help on the canal. Blaine Allphin remembered how surprised he was. "When you dress up to go to church, you expect to go to church. You don't expect to be told to go home and get your work clothes on." They formed a line to pass rocks and sandbags by hand. Ron Hepworth was in charge of their group. They spent all day Sunday filling and hauling sandbags and hauling rock.

"Everybody was afraid to death that the water would get away from us, that we couldn't handle it," said Leland Baker. "Along the Murtaugh canal about four miles was really big-time trouble. We riprapped every inch of that, along both sides. I'd like to know how many hundred thousand tons of rock were hauled in there."

"Sunday, May 20," wrote Julie Randall. "I guess the ox was in the mire. They called us and told us it would only be Sacrament Meeting. Then they called and told us there would be no church at all. I walked over to take some pictures of a caterpillar. It turned out the guy driving it was my cousin. He was working for Hoffbuhr's. They told him to go dig. It turned out to be at our place."

"I can't believe what's happening," wrote Gloria Robins in her journal. "We're sitting here waiting for the dam to overflow. I knew it was going to happen, but until you see fifteen big bulldozers and cats tear your farm apart, it tears your heart apart too.

Things aren't ever going to be the same. We've cried, been angry, experienced every emotion possible. I guess this is the right answer. No one seems to know for sure what is going to happen or how much water we're going to get. It's terrible to sit and wait. We're as ready as we can be, I guess."

Kathy Barrus remembered that their ward was one of the few in Oakley Stake that did not cancel church. "We didn't know when to expect the water, didn't know whether to go to church or not," she said. "We heard the water was a long ways off and decided to go."

The Burley stakes held their conferences: Burley West Stake in the morning, Burley Stake in the afternoon. "Everyone was so tense and excited," one person remembered. "We'd heard they'd turned the water loose, but did not know what it was doing. It was thrilling to think about what had been accomplished, but frightening, too, because we did not know whether the canal would hold."

Usually at stake conference, people were most interested in hearing from the visiting general authority. But at this stake conference they wanted to hear from their local leaders who could tell them about the canal.

"I'm sure all of us are thinking about what is happening a couple of miles west of us, where the water is beginning to come down the canal," said President Petersen. "In the past week we have witnessed two miracles—the marvelous cooperation that made possible the building of a twenty-four-mile canal in three days, and the tempering of the elements—the cool weather that has given us those three days in which to build it."

He talked about people's willingness to work together, about the sacrifice of those who allowed the canal to come through their land. He quoted Churchill: "Never have so many owed so much to so few."

"We need to remember that this project is only half done," he said. "After the crisis is over will come the other half—rebuilding the land and farms of those who have been affected."

Between conferences the stake presidencies, bishops, and emergency directors met with Commissioner Weldon Beck. "Only a few were dressed in suits, since most of the wards had cancelled church services today," wrote Donald Clark. "There was a definite atmosphere of uncertainty as to what the next few hours would bring to our community. . . .

"President Gerratt announced that 5,000 [miners] inches of water had been turned in the Oakley-Burley channel from the reservoir at 8:00 A.M. The Canal Company would increase the flow hourly until it reaches 20,000 inches. . . . The canal had been completed to Blackers, 400 south during the night, as much time as possible is needed to continue cleaning the channel to the river. . . .

"President Gerratt told the Bishops to take charge of their ward area and make the necessary decision in regard for the safety of the people, both member and non-member, they may not be able to contact a member of the Stake Presidency for approval. The Burley Stakes were made responsible for the natural drainage system from 400 South to the river.

"The purpose of this meeting is to organize a 24 hour watch of the new channel banks from the reservoir to the river. Both banks must be watched for erosion and any signs of loss of control of the stream. It would be necessary to send 40,000 inches of water as soon as possible. A major problem would exist because all crossings were out, and crossing the water would be impossible until new bridges could be built.

"Larry Gerratt and Bill Garrard had devised a plan to divide the channel so each Stake and Ward would have their own assigned area. . . .

"President Gerratt emphasized: We must stop the negative talk between County and City. Burley belongs to all of us. We must pull together. Certain people were not being sacrificed for others living in the city. What was being done was the best, least expensive, most logical solution to a major disaster."

At 1:30 P.M. water reached the main street west of Oakley. It reached Randy Hardy's farm at 2:30.

Karlene Hardy watched the water come. "I had no idea I would be so emotional," she later said. "It was not terror, not anger, just an overwhelming feeling at the thought of all that water, at the thought of the power of nature—to think what nature could have done."

At 3:30 P.M. water reached Picketts' gravel pits. "It was a fifteen-foot drop from the canal into the gravel pit," said Ennis

Water approaching Golden Valley road. (Photo courtesy of Nona Jenks.)

Pickett. "I remember when I was visiting the Grand Canyon. I couldn't fathom how that much erosion had taken place just from water until I saw water drop into the gravel pit, and how it eroded the banks for a mile."

"I left the four-mile area and went home," wrote Nona Jenks. "Then I went to Golden Valley to wait for it to get there. It took quite awhile to fill the gravel pits. I talked to people and walked around. It was windy and cold. At 7:55 it reached Golden Valley."

The stake presidencies met at 10:00 P.M. "President Poulton was asked to report on their Burley Stake Conference held today," wrote Donald Clark. "They witnessed improved attendance in spite of men being called from the meeting to help prepare for the flood waters. The talks centered on Charity, Sacrifice, and Love for those who were giving and doing so much for us. . . .

"Burley Stake was assigned to control between 200 and 400 South. The water should be there about midnight. It started from Oakley at 8:00 A.M. A spirit of uncertainty exists since no one knows just what to expect. Fifty men have been sandbagging near the Star Chapel to protect property that may be in danger. Road blocks are set up to control the traffic. Night crews are lined up to

Reservoir water flowing through the diversion gates. The East Canal is at the left, the West Canal at the right, and the old Goose Creek channel in the center. (Photo courtesy of the Walla Walla District, U.S. Army Corps of Engineers.)

watch for any problems, erosion, damage, and etc. They are assigned 4 men per mile at night and 2 per mile during the day. The men will work in different shifts, everyone seems willing and excited to be of any assistance possible."

At the first- and third-lift canals, crews were still installing culverts. "The water was coming at us before we had completed riprapping and backfilling," Evan Rasmussen said.

They cut the 200 West road. Kathy Barrus remembered how tense they all were as they waited for it to be cut. She was afraid it would not be done in time.

"Surveyed the last 2 miles around Blackers," wrote Kent Foster in his field notes. "Assist Blacker beef up his banks around corrals and Hawker with area around his pond. We're going to cut 100 S. road. Called Chevron gas in to locate line."

At his house in Burley, Chevron Pipe Line area operator David Dick was unaware of what was happening. Although an employee of Chevron Pipe Line since 1981, he had been working in Burley for only a short time. "I'd heard people talking about a reservoir that was too full," he later said, "but we were busy trying to get settled in. I didn't understand what was happening, what the fuss was about. When they called me Sunday night, I didn't have the faintest idea what they were doing."

It was dusk when he arrived at the site. Someone from the Army Corps told him they needed to cut through the road. "We have two lines out there," David said. "The gasoline line went through a field so that was no problem. The diesel line was at the edge of the asphalt. We couldn't tell exactly where it was because of the asphalt covering it up. It was about thirty inches deep."

It was between ten and eleven at night. The water was approaching 400 South, about three miles away. They began cutting the road.

"I'd been telling the Corps guy to have the operator raise the bucket on his machine," said David Dick. "I kept saying, 'Raise the bucket, raise the bucket.' The Corps guy was motioning up and down with his flashlight. The operator didn't understand what he meant."

"The diggers were cutting carefully through the road," said Kent Foster, "as carefully as they could cut with one of those big machines. They hit the diesel line. Those big teeth punctured it and it blew all over, covering people with oil. One guy thought he was dead for sure. He would have been if it had been the gas line, but it wasn't. It was the diesel."

"Daylight was almost gone," said David Dick. "In the little glimmer of light that was left, I saw a mist of diesel fuel shoot twenty-five feet in the air. I said, 'Oh, no,' and took off running for my pickup."

The nearest shutoff valve was at Springdale, about five miles east of Burley. As he was driving there, David called headquarters in Salt Lake City on his radio-telephone.

"I told the dispatcher to deliver some fuel to the Amoco terminal in Burley to relieve the pressure on the pipe," he said. "Fortunately, the pipe was not pressured up. There was only about five hundred pounds of pressure. Normally we run about a thousand or two thousand pounds per square inch."

Burley Police Lieutenant Jim Higens was patrolling the area west of the canal near Highway 30. "I live three miles west of Burley," Jim said. "The sheriff's office asked me to patrol over in that area. I was the only law enforcement west of the canal. As I was patrolling, Glen Larson stopped me, told me they had punctured a gas line. I called central dispatch at our command center and headed over there. I was only about three miles away, but by the time I got there, people were already gathering."

"We called Dr. Walter Petersen who's president of the Burley West Stake, and told him we needed volunteers," Jim later told the *South Idaho Press*. "Within 10 minutes there were 50 people there, and within half an hour there must have been 200."

"I was amazed how fast people got there," he later said. "City of Burley flatbed trucks arrived with sandbags. Kids came. Cars converged on the Star Ward parking lot, lined up on the road on both sides. Dr. Petersen was there, lining up people, saying we need sandbags here and here. It was extremely well organized."

It took David Dick about thirty minutes to make the trip to shut off the valves. "When I came back, I didn't recognize the place," David said. "There were so many people."

They had cut through the road and piled dirt around the pipeline. "I told them we had to excavate what they had buried so we could repair it. I happened to have a roll of plastic in the back of my pickup. I was building our house at the time—the plastic was for that. I said, 'We could take this plastic, roll it out, sandbag over to protect it.'

"People and equipment showed up like magic. Somebody showed up with a backhoe. Somebody was there with a truck from Bannock paving. Within a few minutes fifty to a hundred people were there sandbagging. It reminded me of World War II documentaries, of watching the partisans arrive to fight the Nazis."

"They sent out word through the priesthood quorums and over KBAR radio," said Newel Nelson. "It was almost midnight. I had never seen a crew like that come out so fast. Carl from Seven-Eleven had a jug of coffee half drunk. He set it down, grabbed a shovel and away he went. A large percentage who came heard it over the radio. Some came who didn't know anything about it. They just saw all the lights heading that way and followed them."

"Bishop Don Melling had a bunch of high school and college kids in his pickup when he stopped for me," Darrell Roskelley said. "We were out there within about fifteen minutes. I remember President Petersen saying, 'How did you get all these people out here so fast?'

"'Most of them were asleep in my basement,' Don said."

"As I was getting ready for bed, Dad came and told us that a gas pipe had broken and they needed help moving sandbags," wrote sixteen-year-old Scott Melling in his diary. "Shortly after getting there we hopped in a line and started moving sandbags along to be placed up around the hole where the pipe was. We were there an hour or two handing bags of sand to and fro. Some bags were light as a buck and others were heavier than a bale of hay, or so it seemed. Once in a while, they'd get moving the bags so fast you'd barely catch them. Once or twice I got decked pretty good! It was rather fun though. Everyone was working together, and you just had a good feeling."

"Children, adults, old people were out there doing anything they could," said President Petersen. "Those who couldn't lift sandbags held lights so people could see. The esprit de corps was tremendous."

"It was one of the greatest turnouts from the whole community," said Newel Nelson. "People came out dressed however they were. Some kids were there in T shirts. I loaned my jacket to one kid—never did get it back. I didn't know him, he didn't know me—there were so many people out there, he never could have found me afterwards."

"We were wading in diesel fuel knee deep," said David Dick. "Out of nowhere someone from the Burley Water Department arrived with a clamp. Normally it wouldn't work—it's the wrong size—but the cut had distorted the pipe enough that it would fit. We sealed the pipe off with some rubber that I had.

"By the time we got the clamp on, the water was rising. By the time we had finished sandbagging, you could not safely walk across. An hour later there was a torrent of water going by."

Ron Jolley, pipe-line supervisor from Salt Lake City, arrived as they were finishing sandbagging. "I wish I could have had a video camera to take a picture of his face when he got out of his truck and saw all those people and all that equipment," said David Dick. "There were dump trucks, backhoes, loaders, the city of Burley fire truck, all the people sandbagging. I'll never forget his look of dumfounded amazement. 'Where did all these people come from?'" he said.

Meanwhile, the water was continuing to flow down the canal. "It got to our place about 11:30 at night," said Marvin Blacker. "About ten o'clock someone called us, told us to get out. He said, 'There's a mass of water coming your way that will take your home and everything.' We loaded things on our pickup. We had a tent and a canoe. We planned to camp on the nearest hill so we could keep an eye on our house."

"I went in the kitchen, and saw my little four-year-old grandson at the stove," said Lovel. "He was going to waterproof some matches. I'd taught eleven-year-old Scouts for years, and my grandson was often with me. When he heard us talking about the flood, he remembered that in Scouts we'd talked being prepared with waterproof matches."

The water arrived, but not the big wall they anticipated. They stayed in their home.

"After the water crossed Blackers, we flashed lights, watched it cross the road," said Tom Geary. "The whole thing was exciting—to see the enthusiasm of everybody, to see that we could get the job done before the dam overflowed. You work your heart out, and we'd won."

"A lot of us were standing here, watching the water come," said Joan Hawker. "It was kind of exciting. Somebody on a motor bike was playing in front of the water. He almost didn't get out on time before the water caught him."

"Once it hit the corner, it started coming fast," Fred Hawker said. "There was no problem with it staying in the channel."

"When it hit the natural channel then we could see the flow of water," said Weldon Beck. "There was more water than through the third-lift canal. The water roared. Then we realized how bad it could have been."

Bruce Beck was assigned to watch the water from 500 to 400 South. "Bruce called me when it came," said his wife Dawn. "I went to the next mile section to watch. A friend and I were standing there, with the water trickling over our feet. We were like little kids, splashing and playing in the water. I remember thinking, 'It's really here. It really did happen.'"

"We went to bed about 10:30," wrote Julie Randall. "I had just gone to sleep when they called to say the water was at Blackers. We decided to watch it come down."

"The whole family was out there except the babies," Julie said. "We let them sleep. We took the boys—then we really had to watch them. They were so excited. It got to our place about midnight. We could hear it before we could see it. Guys were walking in front of it with flashlights, to make sure it was going right. We didn't get to bed until late that night. People were calling us, we were calling them, to tell them it was here."

"We went to Randalls and followed the water down," said Lane Gochnour. "I remember at Randalls thinking, 'Man, it was like some roaring river, but it was still dry at our house. The water was pushing about as fast as you could walk—kind of trickling in front."

Susan Beck remembered how helpless she felt, waiting for the water to come, knowing that it was going to do damage, knowing that they couldn't stop it. "It's a sacrifice you make for others," she said. "Even though it hurts, you have to do it. We couldn't say, 'You can't come through our land.'"

The equipment was working in front of the water, she remembered. They broke the road just before the water came.

"The water hit about one o'clock," said David Beck. "I was down by Barruses with my flashlight. I knocked on the door, shined my light right in Al's face. I said, 'You've got to get out of here. It's flooding.' I was joking, of course, but I sure scared him for a minute."

At 4:30 A.M. the water reached Harrises' pond. "When the muddy water hit the clear water of the pond, it slipped under the clear water," wrote Randy Harris in his journal. "It made boiling clouds as it arose. There was a distinct line of brown as it moved across the pond." The flood water took several hours to fill the ponds.

It was daylight before the water reached Jay Whittle's pasture. Sherry Gochnour took movies of Ore-Ida trucks hurrying to get across a road before the water came. The water flowed through the culverts under the railroad and Highway 30. At 8:10 A.M. it reached the Snake River.

Water flowing down the new canal.
(Photo courtesy of the Walla Walla District, U.S. Army Corps of Engineers.)

Chapter Twelve

May 21–June 8

At the reservoir, the water lapped against the boards, sixteen inches above the level of the old spillway. It had risen ten inches on Sunday. As more water flowed into the new canal, the water began surging from side to side, eroding the banks.

"Normally you soak up a freshly dug ditch by turning a small amount of water into it," said Ernie Straubhaar. "They couldn't do that. They had to turn a lot in."

"It was a river," said Tom Geary. "It came with force—it was not just a ditch of water. It washed some of our trees and willows out."

"The whole thing was an emotional roller coaster," said Terry Bingham. "I felt good Sunday when the water first went in the canal. Then it started running sideways."

The water would break the banks, thought Terry. People could not haul riprap fast enough. "I was depressed," Terry said. "I thought, 'We're going to ruin all those people who were afraid they would get ruined and we'll have to run it through Burley anyway.'"

Sunday night Terry called National Guard Adjutant General James Brooks. "I told him, 'We need more help or this second canal will get away from us. We're beating the bottom out of every grain-bed truck in two counties hauling riprap, and we still don't have enough.'

"He said, 'You don't need the National Guard, you need a dump truck factory.' Then he told me they had a unit with dump trucks at Blackfoot that he could send."

Lieutenant Larry E. Smith, commander of the 126th Engineer (Dump Truck) Company in Blackfoot, had already been alerted. "On May 16 I received a call from Major Fredrick (Fred) Sermon in Idaho Falls," he said. "He told me there was a problem in Oakley and they might need our unit."

Larry, who had been a commander since the first of May, also taught dental technology at Eastern Idaho Vocation School in Idaho Falls. "The next day in school, I couldn't keep my mind on school work," he said. "When something like that happens, your heart starts beating and you want to go."

Sunday night his unit was activated. "In an emergency like that, when people have to leave work on short notice, they get as many volunteers as they can," Larry said. "They had fifty-one soldiers from various units—twenty-five from my dump truck unit, others from companies stationed at Idaho Falls, Rigby, and St. Anthony. All were experienced truck drivers."

The guard would leave Blackfoot at 8:00 A.M., Terry had been told; but at 8:00 A.M., "here they were rolling into town," he said. "When I saw all those men with all that equipment, I felt like crying. I knew then that we would make it."

"Terry asked us to escort the trucks through town," said Burley Police Lieutenant Jim Higens. "We blocked off the streets. It was so neat to see those trucks—you felt almost like an angel had arrived. It was so great to see someone trying to give you a hand."

Early Monday morning Lawrence Burch was on his way to work when the radio announcer asked for volunteers to come to the city gravel pit. "They said there would be a loader out there. It turned out I was the loader. When I got there, the trucks were lined up four abreast for miles. There must have been fifty farmers out there. I hoped I could get them all loaded before the first ones came back. I was almost finished when I looked up, and there were about forty National Guard trucks coming."

The National Guard had twenty-nine 5-ton trucks and five 20-ton trucks. They worked out of the National Guard armory in Burley with Major Sermon in charge.

National Guard trucks hauling riprap. (Photo courtesy of Randy Harris.)

"Our main mission was to stabilize the banks," said Lieutenant Larry Smith. "We'd haul rocks, shore up banks, head for another place. We were running ragged all the time. Whenever a place would break, they'd need people in a hurry."

"Now the time was spent watching and maintaining a wild river in soft, vulnerable banks," wrote Oakley stake emergency director Wayne Mullen.

A Corps helicopter flew over the canal twice a day looking for trouble spots. Volunteers rode or walked along the banks.

"We divided the canal into three-mile sections," said Rex Gerratt. "Each ward took a section on one side. They were responsible for patrolling the bank twenty-four hours a day watching for erosion. Everyone had a radio. Harold West brought cases and cases of radios. When they found a weak spot in the bank they would call the base station. The base station would contact the National Guard who would then dispatch truckloads of rock to the site."

Reserve Deputy Kent Seale monitored the trouble spots. "If they needed someone to check something out, I'd go," he said. "I had two bosses. Terry would call and say check this. President Gerratt would call and say check that. Finally President Gerratt said, 'We've got to get organized. We can't have you running up and down like that.' Basically I was running around—I was going to say putting out fires, except it was water, not fire."

Kent, Randy Hardy, and Grant Wyatt were assigned to monitor sections of the canal from the dam to 400 South; Burley and Burley West Stakes took care of it from 400 South to the river.

The canal curving around a fertilizer plant and grain storage bins at the Golden Valley road. (Photo courtesy of the Walla Walla District, U.S. Army Corps of Engineers.)

"At first they only wanted three-wheel drive vehicles and motorcycles on the canal—no pickups," Kent said. "They weren't sure how the canal bank would hold. Later they allowed pickups—it was no problem.

"I made a map of my stretch. I'd write on it, 'Has washed five feet at this location,' or whatever. I couldn't remember the small washouts. The canal ran surprisingly well. It wouldn't have been better if they had taken two years to engineer it."

Fred Hawker watched people walk the canal banks. "That was all city people could give, to walk the channel," he said. "They didn't have the big equipment to help in other ways. It was all right. We understood that. They did what they could. I remember seeing President Boyd Poulton—he isn't a small man, but he was out walking the channel too. He stopped and had lunch with me."

"I worried about the people walking the canal," said Kathy Barrus. "The banks were slick. It would have been so easy for someone to slip and fall in."

"It was such an eerie feeling at night," said Richard Hardy, "walking around all alone, looking at water that wasn't supposed to be there."

Although LDS church officials organized the patrols, people of all religious denominations took part. "Several men helped us who were not members," said David Gibbons. "We didn't care whether they came to church, only that they were reliable and strong." The second-lift canal crossed under the 200 West road between 100 and 200 South. The flood waters ran on top of the road. Men raked debris from the headgate to keep it from plugging up.

South of the second-lift canal, the water ponded near Dorothy Lattimer's house. "It covered about sixteen acres of her forty-acre farm," said Alberta Anderson, Dorothy's daughter. "She never complained. When they first talked to her about it, she said, 'If it's going to save going through Burley, that's fine by me.'"

"We assigned home teachers and visiting teachers to look after her," said Rex Gerratt. "She told me later that they took real good care of her. She was so grateful for everyone who helped."

Volunteers sandbagged Dorothy's home and a trailer house, and installed a pump to keep water away from her cesspool. "We checked on that pump night and day," Dave Gibbons said.

Dave was amazed to see how much damage water could cause, how it ate the road bed. It was hard to control the water where it narrowed. "I remember thinking that it was about as wide as Parke Avenue," he said. "When I saw how much difficulty we had controlling that water for a few hundred yards, I knew there was no way we could have controlled it down Parke Avenue."

"Those of us who are not familiar with hydraulics were amazed at the force of the water," said Burley West Stake President Walter Petersen, "whether it came in a single blow, or chipping away slowly. I remember at Harrises. It didn't look bad, but as we watched, within thirty minutes we could see chunks of bank fall into the canal. I gained a great appreciation for the damage water can do."

"They put a lot of riprap along the canal," said Newel Nelson. "It didn't always work. You can't believe how fast that water would cut a bank. I remember at one place the National Guard put three loads of riprap in. Just that quick the water began cutting the other bank."

"The water would wash one bank, then the other bank," said Lawrence Burch. "The stream changed hourly. We'd get one spot fixed—they'd be hollering down the line that another spot needed fixing."

They worked night and day for five days to stabilize the canal banks, remembered National Guard Lieutenant Larry Smith. "We worked around the clock on shifts, but when it was an emergency, we'd all jump and take care of it.

"We had a recovery section. They were trained to pull out a truck that was stuck or about tipped over. Sergeant First Class Benny Billman was head of maintenance and recovery; Kirk R. Bullock was the Warrant Officer in charge. They kept the trucks running. I remember seeing a truck full of rock with two wheels on both sides sunk clear in the mud. The guys got it taken care of."

Volunteers sandbagged the new canal. "Monday, May 21," wrote teenager Scott Melling in his journal. "As we were putting up our new basketball rim, Mom came by and told us they needed help quick, to sandbag. So we changed into our grubby clothes and went out by the same place that we were last night. I had my watering boots on, but I still got them full of water, which was extremely cold. It reminded me of our trip down the white water."

At some places volunteers pounded steel posts in the ground, attached wire to the posts, and piled straw bales behind the wire. The older farmers had suggested using straw, Suzanne Critchfield remembered. It worked well.

At the 400 South road near Burley, Gordon Luke tried to open a plugged culvert. "I had told someone that I had a backhoe," Gordon said. "Somebody called me—I don't remember who. They were concerned about water spreading out so wide and flooding."

The water was 8 to 10 feet deep on the south side of the road, crossing the road at a depth of 1½ to 2 feet. "There was a steel post with a marker saying there was a culvert below it—that's all I had to go on," he said. "That was one scary situation. When water goes across a road it begins cutting from the other side, washing the road as it cuts back. I had no idea whether the water had cut the road. They just said, 'There's that marker there, go do it.' I think I got the job done, but it sure made a guy uneasy. As I was concentrating on opening that culvert, I had the sensation of moving with the water. I remember commenting that it almost made you seasick."

Water flowing over the 400 South road. (Photo courtesy of Nona Jenks.)

Near 200 South and 200 West, water surrounded Martha Lattimer's place on three sides. "The water got here before they was prepared for it," she said. "It was like Grand Central Station with people using the phone. They all had muddy boots on—ruined my rug. They kept talking about a little old lady that was isolated, that they needed to check on. Finally one of them said to me, 'You must be that little old lady that everybody is so worried about.'"

Martha made lemonade and sandwiches for the people working near her place, and "a great big pan of baked beans."

In some places the water cut deep into the earth. "The water didn't come close to ground level here," said Oakley farmer John Martin. It did not touch the rock that lined the canal.

"The water is cutting deep here instead of hitting against the banks," wrote Karlene Hardy in her journal. "I think we filled up the Snake River with our farm."

At the Robins' farm the water was below bank level. "It didn't try to break out very bad," Val said. "As wide as it was, I thought it would be slow. It wasn't. It was like the Snake River."

"It was dirty water—dirty, dirty water," said Gloria.

Evan Rasmussen watched the flood waters flowing over the first- and third-lift canals. "There were no problems with the

Water flowing over the first-lift canal. (Photo courtesy of the Burley Irrigation District.)

siphons," he said. "I was concerned that they would float, like a ship. We had water in them to keep them from floating." He remembered noticing the contrast between the clear irrigation water and the muddy water from the reservoir.

"Monday, May 21," wrote Delma Bean. "The canal looks like a river, being seventy feet wide, and it's muddy water. It looks like chocolate milk running swiftly through our farm."

Viva Warr watched the water in the canal next to their farm. She telephoned her son who lived out of town. "I told him he had to come here to see this," she said. "I told him he needed to come and see that for once Oakley had enough water."

The canal stopped east-west traffic. Burley schools cancelled classes; Ore-Ida closed its potato processing plant. Highway 30 would not be open until crews finished covering the new culverts. Water flowed over the 200 South and 400 South roads, the only roads that had not been cut. Until Monday afternoon when Highway 30 was reopened, people either had to cross through the water, drive to the bridge by the dam, or go on Interstate 15 to Murtaugh.

At the main street west of Oakley, water washed dirt from underneath the concrete bridge abutments. The footings were two feet deep with rock around them, but the water washed deeper.

Volunteers on the temporary bridge at Golden Valley Road, May 22. (Photo courtesy of Donald J. Clark.)

"We had planks stretched across the piers," said Max Lloyd. "People carried rocks and riprap to toss in there to save them."

"People were trying to maintain the piers," said Wayne Mullen. "I remember seeing them start to sink. Max and Gary Lloyd were there. Max said, 'Get off! They're not going to hold.' Everybody got off. They went in just a couple of minutes."

"We were not ready for the water," said Ernie Straubhaar. "Normally we would have got more rock around that bridge. The cement didn't set good enough. If we'd had even a half a day longer we could have got the rock in and let it set and it wouldn't have washed."

At the Golden Valley Road the bridge abutments held. The gravel pits had slowed the water so that it did not wash so deep. Leon Adams arranged to have some large steel "I" beams hauled from Boise to place across the abutments. "All my neighbors helped with the expense," he said. "We borrowed railroad ties and put them on the bridge. Grant Wyatt brought some planking." They finished the bridge Friday, May 25.

Monday afternoon, May 21, the National Guard brought a portable bridge that they placed about five miles north of Oakley. "That bridge was not really for public use," said Karlene Hardy.

"You didn't want to pull on it with a nice car because the entry to it wasn't so great."

On Sunday, May 20, Buzz Bedke and Slim Hewitt had poured forms for a bridge at Cranneys. "They buried 55-gallon barrels for forms, put cement in them, and stuck in railroad iron," Bill Cranney said. "They spanned with nine barrels—had three 12-inch pipe. They set four in the center then spanned from bank to bank." They finished the bridge Tuesday.

At the 100 South road, Chevron Pipe Line repair crews had been at the site of the damaged pipe line since late Sunday night. "By the time the sun came up Monday morning, we had figured out what to do," said David Dick. "We had to repair that pipeline—people from Boise to Pasco, Washington, would have been out of fuel within a week if we did not get it repaired. Because of the way they had cut the road, we had to move that channel of water to the west side of the break. We negotiated with the landowners, hauled pipe in from Salt Lake. Bud Hoffbuhr cut another channel. We were left with a big lake by the pipeline.

"A farmer arrived with two big gasoline-driven water pumps, emptied the lagoon for us. I didn't know who he was or where he had come from. It reminded me of those old Lone Ranger movies—of people saying, 'Who was that masked man? I never got a chance to thank him.' That's the way it was. People would show up, fix something, leave—we wouldn't know who they were, wouldn't even tell them thank you."

It took about two days to repair the pipeline. "I worked fifty-six hours straight—except for a ten-minute nap in my pickup," David said. "We had to replace about a hundred feet of pipe. We had twenty repair people from Salt Lake working on it."

"The company figured they lost $1,700 an hour by not being able to pump diesel," Kent Foster said.

But it could have been much worse, David remembered. "For being an accident, it was amazing how small a problem it was and how effectively it was handled. We lost less than five barrels of fuel, less than 210 gallons. Considering the potential, we were very fortunate. The average person jumped in and went to work."

On Tuesday afternoon, May 22, the reservoir reached its peak—137.53 feet. It was eighteen inches above the old spillway, half-way

The reservoir near its peak, May 21. In the foreground is the spillway with the dam to the left; to the far right is the tower. (Photo courtesy of Ralph Maughan.)

up the second board, exactly where Earl Corless had predicted the peak would be.

Tuesday was Primary election day. Undersheriff Billy Crystal had done no campaigning. "Everyone was working sixteen to twenty hours a day," he said. "Besides the canal, we had our day-to-day law enforcement activities. There were accidents, emergency calls we had to deal with." His lack of campaigning did not hurt him. He won 80 percent of the votes.

Tuesday night Oakley High School graduation was held as scheduled, though the canals caused problems for people trying to attend. The Golden Valley bridge had not been completed, so Leon Adams attached a rope pulley to a small boat. His daughter Terri crossed the canal in the boat to attend her high school graduation.

Wednesday, May 23, Governor John Evans flew to Burley. "It was such a thrill to fly over and see the water flowing down the canal," he said. "Every engineer had agreed it would work, but it had to be done so quick. There was so much apprehension until we saw the water running through and saving the valley."

He talked with people along the route. "Everybody was cheering and smiling," he said. "It was done. We had pulled together and solved the problem. It reminded me of a football game. You didn't have large crowds cheering, but all the way from Oakley to Burley they were cheering individually, smiling and thanking others for their contributions."

"We had a meeting with the governor at Price's Cafe," said Dave Hollingshead. "The reservoir had not increased that day. On Wednesday we were able to equalize the outflow and inflow."

Rex Gerratt remembered what an emotional experience it was for him to see the flood waters flowing to the Snake River. "A radio reporter found me, wanted an interview," he said. "He had me come and sit with him in his van. The water was running. I was all choked up anyway before he came. It was a live broadcast. He got all choked up too. All I could say was, 'We're being blessed.'"

"People were driving out here," said Tom Geary. "It was like a parade of cars going by. They were all coming to see the miracle that had saved their homes and property."

There were willows and other trees where the water came through Gearys' farm. Someone commented that it looked as good as the banks of the Snake River, where so many expensive homes were built. Fourteen-year-old Brad Geary made a sign which they posted near the road: "Riverfront Property for Sale."

"We had a riverside estate for awhile," said Joan Hawker. "Our son, who was a missionary at the time, sent us a letter addressed to Riverfront Estates. It got to us."

Randy Hardy and others near the E-1 Oakley canal irrigated with the flood water. "We had totally overlooked the E-1 canal until a day before they turned the water in," Randy said. "It would be cut by the new canal. Suddenly I realized, 'How am I going to irrigate?'—of all things to worry about during a flood." He helped install a headgate along the side of the canal. It worked, he remembered, "but it was like irrigating with chocolate syrup."

Several people floated the canal in rubber rafts. "People came down the canal in canoes, rafts, every which way," said Jack Gochnour. "Some kids came down in rafts, capsized when they hit our falls, pretty near drowned. It was like the Middle Fork of the Salmon River out here."

With the excitement of the new canal, many people forgot about the canal to Murtaugh. It continued to cause problems;

continued to break out in places. Leland Baker directed volunteers and equipment operators, while the National Guard hauled riprap.

"A group of volunteers tried to repair a bank stretch inaccessible to heavy equipment," wrote *Times News* reporter Hal Bernton on May 21. "The volunteers formed a human chain and dumped rip-rap and sandbags along a water-worn bank. Suddenly they stopped what they were doing and raced to their vehicles to move upstream where the channel had jumped its banks completely."

"Some of the Army Corps' smaller dozers . . . are out [on the Murtaugh channel] with a dragline," Terry Bingham told the *South Idaho Press* May 23. "Numerous farm vehicles and volunteer farmers and the National Guard trucks. . . are hauling riprap. They've been out there shoring up the channel 24 hours a day while the water's just roaring by them."

They could not have contained that water without Leland Baker, Terry later said. "It was his stamina on that canal that kept it from breaking out. The volumes in Cottonwood were great enough that by itself it would have reached Burley."

"One night I was sitting in my pickup by the channel directing trucks to where they should go," Leland said. "First I was getting out and telling them. Then I realized I could stay inside and use my spotlights to point one way or another. The next thing I knew somebody was shaking me. I was pointing all the trucks the same direction. I'd gone to sleep with my hand on the spotlight."

"Leland showed stamina I never knew he had," said Bud Hoffbuhr. "He was there all day and all night every day. He was literally there for the duration. I don't know how he did it. The only sleep he got was for an hour or two in his pickup."

David Beck remembered how exhausted everyone was. Some people had been working nonstop since they first began building the Murtaugh canal. "People's patience got short toward the tail end," said highway manager Ray Kimber. "It was so hard for people to work that many hours, that many days."

"Went to Brother Sagers for income tax material," wrote Ellen Mitchell in her journal May 24. "His eyes looked like two burned holes in a blanket. When we got to President Elquist's home, he and Norma were just coming home from feeding the flood volunteers and he looked and seemed the same."

(Top) David Ritchie, Heather Blacker, and Michael Ritchie by Gearys' sign. (Photo courtesy of Lovel Blacker.)

(Bottom) Leon Adams climbing out of the boat that he and his family used to cross the canal. People on the bank include Alvin Bradshaw, Scott Adams, and Dennis Bergener. (Photo courtesy of Larry Adams.)

(Top) Sheriff's vehicle driving through flood waters on the 200 South road. The flood waters are crossing the second-lift canal. (Photo courtesy of the Burley Irrigation District.)

(Bottom) Oren Gee, Garth Greenwell, and Troy Greenwell floating the canal. "We were the only ones who made it all the way to the river," said Garth. "It was a hair-raising trip." (Photo courtesy of Delma Bean.)

Randy Harris on the sandbags by his house. (Photo courtesy of Randy Harris.)

"I remember on the Murtaugh canal, at ten or eleven at night," Rex Gerratt said. "We'd be so darn tired. They'd call and say they needed so many men. I'd think, 'How can I ask them—how can I ask them to sacrifice more when they've done so much already?' I'd pull over to the side of the road and start calling on the radio. Almost before I'd get the calls made, I'd see a stream of lights heading out there."

The new canal also continued to erode in places. "Brother Mullen reported the bank at 1 mile north of Oakley weakened today and required most of his time but was mended with straw bales," wrote Donald Clark in his minutes for Wednesday, May 23. "Picketts' gravel pit is eroding severely, had taken much of the roadbed for the 2 tracks. The West Canal to Murtaugh needs continual watching 24 hours daily. . . . Nearly lost the water at Merrill Warr's corral where the water is forced from the original channel into the new canal. 25-30 loads of rock seemed to handle the problem."

The water at Picketts' gravel pit eroded the land under the train tracks, carrying away chunks of dirt. Some of the track fell in. Another part was suspended in the air with nothing under it. "We got it stopped before it hit a building," said Kent Searle. "It would have taken the building if we hadn't stopped it."

Railroad tracks shoring up the canal bank at Picketts' gravel pit. (Photo courtesy of Wayne Mullen.)

From inside their house Fred and Joan Hawker could hear the canal water roaring even with the windows closed. They knew it was eroding their top soil.

Jack Gochnour had expected the alfalfa on his land to keep his soil from eroding. "But when they dug to get the pipeline through the road, that started it to wash," he said. "It kept washing, kept moving backwards. We had an eight-foot fall. Where it crossed the road, I put in steel posts and bales of straw to keep it from washing on my upper eighty. That eighty was not hurt so bad.

"The pipeline company was real ornery. They were mad at us because we let them go through the road and bust up their pipe line. It wasn't our fault. We didn't have anything to do with it."

"Every day the falls kept moving back," said Cleo Gochnour. "We knew it was taking out our topsoil. There were days when Jack couldn't go out and look."

"It made Dad really sick," said Lane. "He went right to bed. He couldn't stand seeing all that destruction to his farm, and having people treat him like that. Of course, that bothered us kids."

Near Jack Gochnour's place the water began eroding Shorty Mann's yard. "Then we had a problem," said Rex Gerratt. "Do we rechannel and take more of Jack's land—he'd lost so much already—or do we let it take Shorty's yard? Of course, there's really not

The canal passing between Randalls' dairy barn and houses. (Photo courtesy of the Walla Walla District, U.S. Army Corps of Engineers.)

much choice between a field and a yard. We couldn't let it take out the yard. It was an emotional thing, but Jack was very good about it. I sure appreciated those people that were willing to do that."

Even where erosion was less severe, it was difficult for the farmers. Rex Gerratt remembered seeing 74-year-old Grant Randall on the canal bank. "He was standing there all alone, watching the water flow through his place," Rex said. "His health was not very good anyway. I stopped and talked to him, tried to help him see the big picture. I think he felt better afterwards."

Near Highway 30 the water eroded Harrises' back corral. "The canal went around their house, barn, corrals—the works," said Newel Nelson. "It came across the road going east, turned north and went through their corrals, then cut back west and made a circle around their home and barnyard. It cut deep channels."

"There is a nine foot drop every quarter of a mile on the last part of the channel," said Ed Harris. "The channel was like the white water on the Salmon River, only it was mud water."

"At night we could hear the waterfall," said Marian. "It would break off land as it came up. It washed and washed, got deeper as the

Water northeast of Harrises' home. (Photo courtesy of Rachel Gorringe.)

weeks progressed. The waterfall was moving, cutting dirt away. It moved up and back and up and back."

"The water falling would not bother our house because we were high," said Ed. "But it could have taken Randy and Denise's trailer. They were all packed up, ready to move."

Dave Gibbons helped sandbag at Harrises. "It was kind of fun," he said. "We were cracking jokes and tossing sandbags. I remember thinking it was almost a sacrilege to be laughing and joking like that when there was so much tragedy and potential tragedy for those people."

The National Guard hauled big rocks from Declo to try to stop the erosion, Lawrence Burch remembered. "One rock was about five feet in size. When they dropped it in the canal you couldn't see it, but you could hear it—clunk, clunk, clunk downstream as the water washed it away."

Someone from the army talked to Ed about what they could do, Ed remembered. "I looked at the double fall and the way it was traveling and told them to sandbag it. The sandbagging held three days. Then the waterfall took off again."

(Top) Water in the old Goose Creek channel entering the man-made channel at Merrill Warr's place. (Photo courtesy of Steven Pearson.)

(Bottom) Water crossing the 200 South road near 200 West. (Photo courtesy of Steven Pearson.)

(Top) Flood waters covering Jay Whittle's pasture. To the far right is Lane Gochnour's dairy, the trees almost covering the barn. (Photo courtesy of Nona Jenks.)

(Bottom) Water ponding around some old chicken coops on Harlan Jensen's property. (Photo courtesy of the Walla Walla District, U.S. Army Corps of Engineers.)

It was a difficult time for the Harris family. "Carolyn Hanks and I always had our heads together," said Marian. "It helped to talk to each other. Other people didn't understand what we were feeling. Sometimes in the middle of the night I couldn't sleep, so I'd go out walking. I'd bump into my brother David, who was also walking. He couldn't sleep either.

"A few thoughtless people had a lot to say," she remembered. "People from town would come out, look at our place, and say they didn't think it was such a big deal. We weren't being hurt any. They didn't know. They hadn't seen our place before. They couldn't see the damage. They didn't know what we were feeling."

People in Burley were talking about the canal, Kathy Barrus remembered. "A couple of ladies commented that they didn't think it was such a big deal to go through farmland. It made me feel so sad to hear them talk that way. Damage to crops and land is no different than damaging someone's store, taking out their inventory. Those people were my friends. I knew what they were going through, how hard it was for them. It made me sad that people were not more appreciative."

Gloria Robins remembered that someone who was walking the canal told her son, "'Us city folks can't do this any more. We have to go to work. You'll have to do this for yourselves.'

"I wondered what he thought we did," said Gloria, "sat around all day and napped? It made me realize that you never know how other people live and feel if you're not part of that group. People in town don't know what farmers do and vice versa."

But many people in town were grateful, she also remembered. Some people wrote letters to the editor expressing their thanks.

"We got a card from Juanita Kerbs," said Gloria. "She said how sad she was for us, how bad she felt, wanted to know if there was anything she could do. We got several cards from people—sympathy cards, I guess you'd call them."

Price's Cafe sent a gift certificate for a dinner. "They sent one to everybody along the route, thanking them for letting the water come through their place," said Martha Lattimer. "I wasn't sure about it, so I called up. They said I could come in any time and bring someone with me, and have anything I wanted to eat, we wouldn't have to pay for it. So my son and I went in and had a steak."

The McNeil-Lehrer NewsHour aired Wednesday night. "I was a little disappointed," wrote Nona Jenks, "because they never mentioned Oakley. All it was, was Burley. I didn't think it did justice."

People had expected a report about the canal, but only 3½ minutes of the ten-minute report described the canal and threat from the reservoir. The rest focused on Marvin and Lovel Blacker.

A PBS camera crew had been at the Blackers' farm for several days. "They were here from morning until night," said Lovel. "They even had their meals with us. They took so many pictures—of people sandbagging, bringing us food—they must have had twenty hours of film. They followed Marvin around constantly. We became well acquainted with Patti Parson who was over the camera crew. She wrote to us for a couple of years afterwards."

That the PBS crew sympathized with the Blackers was evident from the resulting newscast. "It's the story of a few people having to make a sacrifice to help the rest of the town," Jim Lehrer said.

That sacrifice became the theme of the program, with the visuals, Lehrer's commentary, and quotes from Blackers chosen to show that in this flood, Blackers were the victims. The report helped viewers understand the problems which Blackers and the other landowners faced as it graphically portrayed the landowners' sacrifice, which was a part of the building of the canals.

But it was only a part. Because the McNeil-Lehrer NewsHour portrayed it as the whole, it distorted the experience by taking it out of context. Although the program was broadcast Wednesday, May 23, most of what was shown had been filmed Friday and Saturday, May 18 and 19, during the time of greatest uncertainty for the landowners, as they waited for and then watched the bulldozers dig through their land.

"The canal is cutting the heart out of our farm, and I feel like it's cutting the heart out of my husband," said Lovel Blacker. "From boyhood his dream was to have a farm of his own and a family on a farm."

"I guess this is a world of give and take," said Marvin. "Sometimes it seems like it's the same people that take and the same people that give and I guess this is right. In the end it will all come out, we hope. I hope I'm big enough to make it through."

Marvin and Lovel did not say anything different from what other landowners were saying during that time of stress. But the

Marvin Blacker talking with SCS technician Burke Scholer. (Photo courtesy of Lovel Blacker.)

effect when shown on television was different. What other people said would be forgotten; what the Blackers said was preserved. The visuals showed their worried faces, showed how it hurt them to watch their land being destroyed.

Jim Lehrer's comments contained inaccuracies which added to the impression that the Blackers were victims. "His irrigation system will be disrupted, and he will have to travel six miles to get to his wheat crop because he won't be 'allowed' to build a bridge over the canal. . . . If the banks of the new canal don't hold, the Blackers will have even more problems. Last week the banks of another city-built canal eroded and finally caved in. . . . The canal had made it to the Blacker's farm before the dam overflowed, but they had lost almost twice as much of their land as promised."

The report said nothing about how long and hard people had worked on the Murtaugh canal, did not explain why the Snake River canal had to be constructed in three days, and barely touched on the damage that would have occurred had the canals not been built. It did not show people helping Blackers. Except for a ten-second spot of Marvin and his daughter sandbagging, it did not show Blackers helping themselves. Most of the scenes were of Marvin and Lovel waiting and watching.

"I was surprised to see the program," Lovel later said. "They asked us so many things—they were days filming here—and then they put in so little."

"[Marvin] was surprised to learn that a televised statement might sound a little different than it was intended after the tape has been edited to save time," wrote reporter Mike Hill in the *South Idaho Press.*

Local people watching the broadcast had ambivalent feelings. While most recognized that Blackers' concerns were legitimate, they did not like the subtle undertone of accusation. If Blackers were victims, then the local people were responsible, because their lands had been saved. People did not like being made to feel guilty. They did not analyze the report, did not understand how much editing had gone into it, nor how that editing had shaped the mood. Most had only a slightly uneasy feeling, as if something were not quite right.

The timing of the broadcast added to this feeling. The program was filmed during a period of uncertainty and anxiety, when the farmers had to watch the bulldozers tearing up their lands. It was shown during a time of rejoicing, when everyone knew that the canals had saved the valley. People wanted to savor the victory, not be reminded of the casualities of battle.

The report did not tell the main story of the canals; that story was too big for anyone at the time to comprehend. It did not tell the main story of Blackers and the other landowners even though it portrayed their plight with great empathy. For in spite of not liking it, and in spite of having difficulties with their feelings when they had been asked to sacrifice, Marvin and Lovel had allowed the canal to come; they had not tried to stop it. They had honestly expressed their feelings, but they had not let those feelings keep them from doing what had to be done.

Marvin was a member of the Oakley Stake High Council. Earlier he had worked with President Gerratt and others who had pushed for the building of the canal. He remained on the High Council and continued to work with President Gerratt and the others. "I don't know of a finer person than Marvin Blacker," President Gerratt said.

"It's a learning experience for us and I guess we'll grow by it," Marvin said at the end of the McNeil-Lehrer NewsHour. "We're going to have a few scars and stuff. It will take awhile to get over some of the feelings we have, but we'll make it." He and the other landowners would make it. And that was the real story.

After the broadcast, Marvin and Lovel received several letters from people all over the United States. "They were addressed to Blackers, Burley, Idaho," Marvin remembered, "but they got to us. One person asked, 'Did they really build a canal in three days?'"

On May 25 the National Guard had been scheduled to go to Boise for annual training. Governor Evans asked that the Guard train in Cassia County and continue working on the canal.

"The National Guard has a two-fold mission," said Lieutenant Larry Smith. "We're responsible to the governor for state emergencies, and to the federal government for national defense. The annual training is to prepare for our wartime mission. Brigadier General Lawrence B. Campbell called me. He said, 'Do you think this would benefit your unit to remain in the area, and would not interfere with your wartime training mission?'

"I said, 'Sir, this is the best training our unit could have.' Our wartime mission is to pick up and move materials, to haul things for whoever wants them. I said, 'We're in an emergency situation working twenty-four hours a day. We're doing in a real live mission exactly what we'd do in war except that no one is shooting at us.'"

On May 26 the full 126th National Guard Unit arrived in Burley. Major Sermon left; Lieutenant Smith became commander in charge. He moved his unit to BLM land between the two canals.

"In my eleven years' experience in the military, I've never seen a unit with higher morale," said Terry Bingham. "They were so good to work with and so eager to help. Without them we would not have been successful."

While the Guard was working, "the community was hauling food to my soldiers," said Lieutenant Smith. "They thought it was great. The only bad thing for me was that we have cooks in our unit and meals ready to eat. My food count was way off at first."

The reservoir level, which had peaked at 137.53 feet on May 22, had decreased slightly. By Friday, May 25, it was 137.21 feet. Dave Hollingshead had been releasing 810 cfs into the new canal since

Wednesday. An additional 239 cfs flowed from the reservoir into the West Canal and Murtaugh Lake; 39 cfs flowed into the East Canal. Because the East Canal had been filled with water from Birch Creek, no reservoir water had been diverted into it earlier.

The weather service had predicted that because of warmer weather, a new peak would hit the reservoir on the weekend. On Saturday, May 26, Earl Corless and Leo Busch flew over the Goose Creek drainage. "Our purpose was to see how much snow was left in the mountains," Earl said. "We checked soil moisture to see if the snow would soak into the ground. We felt comfortable at that point that the major peak had gone."

But even though the danger was largely over, the inconvenience remained. "My house was on one side of the channel, and most of my farm was on the other," said Fred Hawker. He and others had been driving through the water at 400 South.

"Then one day I started driving and felt the end giving way," he said. "I got out of there and set up some bales of straw to stop traffic. After that I had to travel eight miles to get to my field."

"Our potatoes, eighty acres, were on the west side of the canal," said Val Robins. "I had to get them sprayed—had to go to Ore-Ida and back to do it. I tried to throw wrenches across the canal to my boy who was cultivating. It was too wide."

Val's boys had to cross the canal whenever they changed water. One day, "instead of going that long way around, Bret decided to put his younger brother in a tank—an old wooden trough—and let him float across to the other side," Gloria said. "He got the tank and a rope, put his brother in it. Then they realized it wouldn't float to the other side, the water was too swift. It started to sink, and Joel got out of there quick."

The Robins' neighbors had an irrigation pipe over the top of the canal. "I sent some pipe movers to shimmy across," said Val. "They came back and said, 'We don't want to go that way any more.'

"I said, 'It's not so bad.' So I went across it, and then I could see what they were saying. You get twenty feet out and have seventy feet to go and that water is raging beneath you like a river—it's a scary deal."

"I had to drive seven miles to get to my pump," said Randy Hardy. "We had a ten-inch irrigation pipe across the canal. I was scared of walking across it—I'm not one who takes to that kind of thing. At first I'd shimmy across, but it was an old pipe, wrapped, with a lot of tar. It made a mess of my pants. Finally I ignored my feelings and walked across. You'd get to the middle of that seventy-foot span and it would start bouncing. It bounced with every step you took."

"I'd be holding my breath when he'd walk across that pipe," said Karlene. "I hated to have him change water."

Near Highway 30, Lane Gochnour's driveway was the only access to several homes. "I was so worried about my daughter Jolene," said Marian Harris. "She had twins, was pregnant at the time. Whenever she went out she had to cross a fence."

They did not cut the road near his house, Lane Gochnour remembered. "It already had a culvert underneath it. It was a box flume—sides and a top, but no bottom. It held for a couple of days. President Gerratt came by in his pickup, parked right on top of it. Then he said, 'I better move my pickup.' He moved it, and a couple of seconds later the flume fell in."

Carla Gochnour was taking pictures of the canal with a movie camera and caught the whole thing on film. It showed the pickup driving off and the road caving in almost simultaneously.

At first most traffic went along a little road below Lane Gochnour's corral. "Then one night about 2:00 A.M. someone from the National Guard called," said Lane. "He asked me to move my pickup—it was parked right by the house. The road by the corral had become too soggy for trucks. They started coming up by the house."

"I was sure glad the kids were at Soda Springs with all those trucks coming by," said Carla. "It was like a freeway out there."

The Randalls worried about their children falling into the canal. "The scariest part was having that dangerous thing in our yard," said Julie.

Rosanne was concerned about her youngest child, who was eighteen months old. "She and I stayed in the house most of the time," Rosanne said. "I didn't dare let her outside. The canal was built right through her play area."

The first week was "really chaotic," remembered Val Robins, "but we got used to the routine after awhile."

By the end of the week, the Murtaugh diversion canal was finally under control. With riprap lining the banks and flows from Cottonwood decreasing, the danger from breakouts was finally over. "They kept battling it and won," said Terry Bingham.

"Sunday, May 27, went to church," wrote Carl Van Tassell in his journal. "We'd rather go to church than haul rock."

"I'd like to give you the dam report," announced Bishop Don Melling to his congregation on Sunday. He paused a minute, then added, "I've always wanted to say that in church."

Sunday evening at Pella the first of a series of landowners meetings was held. "The meeting was directed by the three Stake Presidents from Oakley, Burley, and Burley West Stakes, with President Rex Gerratt presiding and conducting," wrote Donald Clark. "Pres. Gerratt explained the Reservoir had been dropping about an inch or two daily since Wednesday. . . . The water is currently below the boards but still about a foot above the old spillway. Without the work of the last few weeks we would definitely be flooded. . . . Twenty-four hour vigilance over the canal will continue."

They opened the meeting for questions.

"We hear the channel will stay for two years, please comment."

"Can any precaution be taken about washed out trees lodging in the culvert at Highway 30?"

"Places where silt is being dropped, can it be farmed this year or when?"

"Could the Army be advised not to run over my portable mainline? It has happened twice now."

"What can be done about bridges across the channel?"

"Can you estimate the cost of a bridge?"

"Answer: From $0 to $15,000 depending on ingenuity and the help of individuals and businesses."

"How is the rock holding out?"

"Answer: We still have some, but I cringe when I think about hauling it out."

Marian Harris later remembered how helpful the meeting was. "They had meetings for people to relieve their feelings, to talk about it," she said. "President Gerratt kept our nervous system intact."

"They were trying to encourage us and thank us," Val Robins said. "It was hard to be encouraged." But he also remembered that most people were pretty calm. "A few things were said that shouldn't have been. After it was over, it was forgotten. I don't think anybody hated anybody. There were no feuds. The people in Burley were trying to do everything they could."

Bart Bowers remembered thinking that as hard as the people from town were trying, "They just didn't savvy. They tried to understand, but the guys in town never had anything to do with earth or equipment. President Gerratt—there was nothing he didn't feel or understand. It really laid on his shoulders."

Later, the stake presidencies discussed the problems raised, particularly the need for more bridges. The canal company was arranging for a bridge on Oakley's Main Street; David Beck and Randy Hardy built bridges at their farms.

"It was like Christmas when we got that bridge done, when I was able to get back and forth without walking that pipe," Randy said.

"The Burley Stake accepted the challenge to place a bridge at Randalls," wrote Donald Clark in his minutes for May 27. "The Oakley Stake will build a bridge at 600 or 700 South."

Earlier someone had volunteered to build them a footbridge, Rosanne Randall remembered, "but I didn't want it because of the children. I was afraid one of them would get on there and fall off. But one day Richard almost had an accident because he was so tired from driving all that long way—12½ miles to get to our barn—so we told them to come and put one in."

Grant Wyatt helped Burley Stake construct a bridge. "We borrowed two long poles from Unity Power," said Tex Woolstenhulme. "I bought a bunch of plywood. We put a floor on it and sides for the kids."

"The kids liked that bridge," said Julie Randall. "It was fun for them to have. They couldn't see over the plywood sides. They liked to run across it and make a racket. It creaked and swayed. I was afraid it would break, but it didn't."

Max Lloyd located an old single-lane bridge at Declo that had once spanned the Snake River. The canal company purchased it. "It was a big, heavy, steel bridge," said Winslow Whiteley. "It took two cranes from Twin Falls to move it."

Tuesday, about noon, Rex Gerratt made arrangements for the National Guard to bring a heavy wrecker truck to 700 South at

5:30 P.M. to lift bridge material off a semi, "not knowing at the time what the bridge was going to be or who was going to haul it," he said. "Then I went looking for one. I saw Grant Wyatt, said, 'Jump in, we're going to get a bridge.'

"You don't find bridges lying around the main streets of Burley, so we headed for Twin Falls. You don't find bridges along the main streets of Twin Falls either, so we began looking in people's salvage yards. Behind Idaho Frozen Foods we saw two beams seventy-feet long welded together with pipe. I went in and asked for them. The guy didn't know me, but all he said was, 'If you'll sign a slip for these beams and bring them back when you're done, you can have them.' I told him we'd have a truck there in an hour to move them.

"We needed a truck to haul them to Burley. We went to Twin Falls Truck Freightliner dealer Cotton Crumbliss whom I knew. I said, 'Cotton, I need a truck to haul some beams to meet the National Guard wreckers near Burley at 5:30.'

"He said, 'We don't have a truck here.'

"I said, 'Can you get one?' He said he could.

"Then we went to Northwest Crane and Rigging. I told them to go to Frozen Foods to load two beams on a truck that would be coming. They'd never seen me before, but said they'd do it. I drove back to Burley, dropped Grant Wyatt off, and drove to the bridge site.

"At 5:20, here come the National Guard wreckers from one direction, and there come the truck from Twin Falls with the over-length beams from the other direction. Barclay's and Wes's welding were there to cut off the pipes.

"I sat down and cried. People were so good. That's the way it was throughout the entire project. It didn't matter who it was or whether they knew us or not. Everybody was so willing to help."

May 28 the Corps of Engineers left Cassia County. Colonel Robert Williams, commander of the Corps, toured the area the last day. "Everything about this job has been unusual," he told the *South Idaho Press.* "Both the cooperation and the participation of local people has been phenomenal."

J.J. McLaws collected his $50 bet. "I bet that the water would not run over the spillway," he said, "and it did not run over."

(Top) Bridge at Cranneys. (Photo courtesy of the South Idaho Press.*)*

(Middle) Iron bridge at Oakley Main Street, with the washed-out concrete piers underneath. (Photo courtesy of Donald J. Clark.)

(Bottom) Beans' irrigation pipe across the canal. Farmers used this and other irrigation pipe as a footbridge. (Photo courtesy of Nona Jenks.)

(Top) Bridge at Golden Valley. (Photo courtesy of Jim Mai.)

(Middle) David Beck holding his son Matthew near the bridge he built at the 300 South road near his farm. (Photo courtesy of Susan Beck.)

(Bottom) Footbridge at Randalls' dairy. To the right is their cellar. (Photo courtesy of Richard Randall.)

But water had run over the spillway, although few people knew it. In the middle of the night someone had siphoned a little water over the top. Whoever did that had a bet he did not want to lose.

Harold West returned the radios to the Boise Interagency Fire Center. "As soon as I had a full kit, I'd turn it in," he said. "At the end there was one radio missing. I stalled as long as I could trying to find it. I finally sent the kit back minus the radio. I was personally responsible—technically they could have made me pay. They were worth about $3,000 apiece. I asked Channel 11 TV if they could do something about it. At first they said no. Then they called up and said they'd do a human interest story on it. On TV I said, 'Look again. Maybe it slid down behind your seat or something.' A few days later someone turned it in to Terry."

The reservoir level continued to decline. By May 27 it was off the boards; by May 31 it was below the spillway addition.

After patrolling the canal for ten days, the volunteers stopped walking and riding along the canal. "I continued to ride it after they were done," Kent Searle said. "I'd go three times a day, then two, and finally only once a day.

"I remember one night, my brother-in-law was visiting. I said, 'How would you like to go for a nice ride on the canal bank?' It was about nine o'clock. One of the farmers had not shut off the end gun of his pivot sprinkler. I got stuck. The dirt was so greasy and slick. Instead of a quiet ride, I had a two-hour mud fight. Before I was done, I had three- and four-wheel-drive vehicles and pickups all out there trying to help me. They all got stuck at least once. It was midnight before we all got out. I was a little put out at that farmer. Here we were guarding his field, and he didn't shut off his end gun."

There were still some problems. "The water in the channel changed direction about ¾ mile south of Oakley when it tried to seek the old channel today," wrote Donald Clark on June 4. The National Guard continued hauling rock.

The erosion at Harrises grew worse. "The steep water kept unravelling and washing back to some houses," said Burke Scholer. "They were running rocks and dumping them in. The water had so much velocity. It washed the rocks and kept eroding. Terry called us at 6:00 P.M. Sunday. They were having such a big problem—wanted our help. We had the Guard trucks haul huge rock from near Declo. We stockpiled all the rocks, and then pushed them in all at once with Kloepfer's big loader."

"One rock was so big it wouldn't go out the end of a twenty-ton truck," said Ed. "They had to lift it out."

"Those were huge rocks," said Terry Bingham. "When the trucks dumped those rocks, the ground shook."

The rocks finally stopped the washing. The canal held until the flow of water was cut.

The National Guardsmen were scheduled to leave the area June 9. Price's Cafe furnished dinner for them at the Burley Armory, and city and county officials expressed their thanks. Wade Baker, who had a swimming pool on a natural hot spring near Oakley, hosted a swimming party. "He opened up his pool to the whole company," said Lieutenant Smith. "He had a bunch of food—everybody had a great time."

After studying the conditions that had led to the flood, government officials declared it was a 100-year event; a problem of that magnitude could be expected to happen once in 100 years. Later, after collecting more data, they revised their estimate and stated that it was a 200-year event.

With the inflow from Goose and Trapper creeks declining and the reservoir level continuing to decrease, officials talked about how soon they could turn water out of the Snake River diversion canal. Flows from Cottonwood had decreased enough that the Murtaugh diversion canal and the regular irrigation system could handle the surplus water.

Friday, June 8, the diversion gates to Goose Creek were closed. "The rickety old structure that had creaked and groaned and threatened to give up on us several times was dammed off," said Terry Bingham. The flood water ceased flowing to the Snake River.

"What a relief it was when they turned the water out and we could run our motorcycles through the canal," said Val Robins. "It made it part of the farm again."

David Beck in the dried-up canal.
(Photo courtesy of Susan Beck.)

Chapter Thirteen

June–December

Monday, June 11, three days after the water stopped flowing, Rex Gerratt went to Salt Lake City for knee surgery. Although there had been several accidents—trucks and backhoes tipping over, one or two persons falling in the canal—his was the only injury requiring treatment. In spite of being in pain for three weeks, Rex had kept going until the job was done.

Tom Tate had not quit either, though he knew he was not well. "I knew there was something wrong because I was so tired," Tom later said. "Of course, I wasn't getting much sleep, but it wasn't that. It was a different kind of tiredness."

He continued to work without stopping or slowing down. He did not go to a doctor until he returned to Walla Walla.

"Then they found I had multiple sclerosis," he said. "They figured I'd had it for a long time, but didn't know it. My body had always been able to fight it before—in twenty-five years I'd never had an hour of sick leave. But the intensity of that experience in Burley was too much. My body couldn't take it—couldn't fight it any more."

Building the canals was Tom's last major project. In October, 1984, he was forced to retire on a medical disability.

Few people in Cassia County were aware of Tom's problem. They were now facing the task of assessing the loss, restoring fences, repairing damaged farmland and disrupted lives.

Hawkers' dog Sassy in the nearly dry canal at their farm. (Photo courtesy of Fred Hawker.)

"Harder than seeing it come was seeing it go," said Joan Hawker. "We didn't want to see all the destruction, didn't want to see what it was going to leave."

The water washed away all the fill dirt that had been brought in through the years. While the canal was being built, Fred had eight inches of his top soil pushed up, to save it. Now he regretted not having saved much more.

"It raped the earth down to bedrock," he said. "At one place the water made a hole like you would drill with an auger—about three feet deep, eighteen inches across, straight down, pretty as a picture, just as round."

Another field was silted in and covered with rocks. "It looked like Craters of the Moon," he said. "We were picking up garbage that had come all the way from Oakley. We found one antique license plate. The silt from Oakley was like quicksand. It wouldn't dry out. You couldn't walk on it. Later I ruined some equipment trying to farm it."

There was silt in places at Randalls. "We couldn't farm it," said Richard. "It was so spongy and soft we didn't dare get on it. The silt crusted over on top, but didn't dry underneath. It was like jelly—not solid."

Rocks at the Hawker farm. (Photo courtesy of Fred Hawker.)

Part of Martha Lattimer's pasture land had been low, "so low that cows would go out of sight down there, so that you couldn't see them from the house," she said. "The flood washed silt and debris down there and left it. It raised it even with the other ground. It looked like it was dry on top, but it was like a marshmallow. We couldn't farm it for two or three years. Couldn't get a tractor on it."

When the field by Barruses' house was planted the following year, "everything grew except where the water had been," Kathy said. "You could see exactly where the water was, because there were no crops there."

The water spread wide at Bowers' place, "but it didn't move any dirt," said Bart. "We received real estate instead of having real estate wash away." Rocks had washed into his field. He also found cockleburrs. "We never had cockleburrs before," he said. "The year after the flood I hauled out over a pickup-load full."

The flood left lava ash, weeds, and willows on Weldon Beck's farm. On the Robins' farm, "wild oats came up the next year just terrible," said Val. And year after year the farmers picked up rock.

"We couldn't put the fences back because of the rock," said Marian Harris. "We have tons and tons of rock—but we didn't want rock."

(Top) Paula and Karla Watterson under the bridge on Oakley Main Street. Behind them is a concrete pier. (Photo courtesy of Paula Watterson.)

(Bottom) The DeLayne Anderson family in the diversion canal near their home. (Photo courtesy of Randy Harris.)

(Top) Sandbags and rocks lining the canal at Hawkers. His irrigation pipe is above the canal. (Photo courtesy of Fred Hawker.)

(Bottom) Silt covering a field. (Photo courtesy of Ryan Pearson.)

Jennie Hale, Kristen, Jaclyn, and Ben Hardy on Randy Hardy's bridge. A pool of water still remains in the canal. (Photo courtesy of Randy Hardy.)

The water eroded the soil eight to ten feet deep in Jack Gochnour's hayfield. "I had somebody from the ASCS come out and figure the cost to put it back the way it was," Jack said. "He told me it would cost $80,000. I didn't have that kind of money." When Jack tried to drive his tractor near where the canal had been, "it would sink to China."

In spite of his loss, Jack did not question the decision to build the canal. "It was the way to do it," he said. "It would have cost billions of dollars in damage if they hadn't done it—would have taken out half the town."

From May 20 through June 7, approximately 37,500 acre feet of water had been dispersed from the reservoir—almost twice as much water as was in the reservoir at any time during the 1930s. The East Canal had handled approximately 2,200* acre feet; the West Canal and diversion to Murtaugh, 9,700 acre feet; the diver-

* The total flow and the flow in each canal were measured separately. The figures for the combined flow from the three canals are not the same as the total from the reservoir because of the imprecision of measuring streamflow.

sion canal to the Snake River 25,400 acre feet. The Murtaugh diversion canal had also handled an unknown amount of water from Cottonwood Creek—perhaps 10,000–15,000 acre feet—the East Canal unmeasured water from Birch Creek. An additional 6,638 acre feet had been dispersed from May 1 through May 19.

No one had anticipated a diversion of that magnitude. On May 1 the SCS had predicted a May-July inflow of 47,500 acre feet—240 percent of average. The actual May–July inflow was approximately 77,600 acre feet—392 percent of average, with 47,100 feet flowing into the reservoir in May. The runoff for the entire water year was approximately 119,700 acre feet; the carryover from the previous year was 34,600 acre feet. Thus the total amount of water was 154,300 acre feet—more than enough to fill two Oakley reservoirs.

Most people, however, did not understand what had been accomplished. Until May 15 only a few Burley residents had realized that their town was in danger. Then, before they had time to comprehend the seriousness of the situation, the diversion canal was built and the flood averted. Most people had never experienced serious flooding. Those who knew little about water did not understand how much damage water could do. To avert panic, the government agencies had purposely played down the emergency.

"People never realized the danger they were in," said Earl Corless. "People do not know about floods that never happen. It's human nature to believe that a disaster will never happen to you."

"A few people said, 'That whole thing wasn't really necessary,'" Larry Gerratt remembered. "It was hard for me to imagine that people could see all that water running for three weeks and not realize how much damage there could have been. There were not very many who felt that way—only one or two. But I was surprised that there were any."

The people who knew water, the professionals who dealt with it scientifically and the farmers who worked with it daily, understood what would have happened. Kent Foster estimated that if it had not been for the diversion canals, there would have been $60 million damage to farmland alone, not counting damage to Burley. "Some people would have been folded up completely," he said.

In addition, the BID canals would have been ruined, leaving 20,000 acres of cropland without water. "We would have lost the agricultural tax base of the county," said Commissioner Weldon Beck.

Those near Oakley would not have been hurt as much as the people downstream. "It could have been a catastrophe to the city of Burley," said hydrologist Leo Busch. "The damage would have been immense. It would have spread out across the landscape. People don't realize how much it would have spread out."

"The water was running fast," Bob Sagers said. "It would have cut up a lot of farms. It would have spread out, then gravitated back to the channel, then spread out again. It might have taken three days to get to Burley, but it would have reached there. Outside of the channel there might have been only about two feet of water, but that's six inches above the sidewalk—enough to get in everybody's houses, fill everybody's basements."

"Water would have been running in Burley for weeks," said Leland Baker. "It would have gone into the sewer system and contaminated wells."

"The storm drains would have filled with mud," said Olene Spurgeon. "We'd still be trying to clean storm drains today."

"The damage to Burley would have been in the millions," said engineer Dal Wayment. "There would have been significant damage to property value. All FHA loans would have been re-evaluated."

There would have been flooding from Cottonwood if the canal to Murtaugh had not been built. The farmers along the Cottonwood drain, where the second canal was built, would probably have been flooded. Cottonwood water would have come without warning, giving people no time to prepare.

Later Jason Allen made a map for his Eagle Scout project using SCS figures to show where the flood waters could have gone and how long it would have taken to reach each location. His map showed that the waters could have gone anywhere in Burley. "The area is so flat that you'd have water backing up and changing direction every time it hit a haystack and every place where trash and debris was pushed up against a fence," Kent Foster told the *South Idaho Press*.

"I'll tell you where the water would have gone," said Bill Cranney. "It would have gone to wherever the person with the biggest machinery could shove it. Whoever had the biggest D–8 cat to shove it over to their neighbors wouldn't have got hurt."

"Everybody wonders what it would have been like if the canals had not been built," said Newel Nelson. "After working in Rexburg, I'm grateful we weren't cleaning mud like that out of our basement."

"We built about 42 miles of canal system that normally would have taken years to complete in just a few weeks, and to my knowledge, not one home is under water," Rex Gerratt told the *South Idaho Press* on May 27. "From the damage standpoint, I feel like the first day those canals were opened, the bill for this was paid in terms of property that was saved."

But few of the affected landowners were thinking about that as they watched the canal dry up. They were thinking about covering the canal and getting their lives back to normal.

The commissioners advised them not to fill in the canal. "We don't even know who owns the canal," Weldon Beck told the *South Idaho Press* on June 12. "We have to find out where we are legally."

But the farmers did not wait. "Raymond said, 'I'm not going to leave that there,'" Delma Bean remembered. "Weeds would grow up and make a mess."

A construction company gave Val Robins an estimate for covering the channel. Ray found he could rent a new D–8 cat for less.

"Tuesday, June 19," wrote Delma in her journal. "The D–8 caterpillar came and Bruce got on it and really learned to run it fast. It's a wonderful machine."

Beans filled in the canal at their farm, then began filling it in for their neighbors, most of whom shared the expense. It took a month to cover the 5½ miles of canal from Adams through Blackers. The cost of filling it in was estimated to be about $10,000 per mile, which included $3,800 for the cat, plus fuel, labor, and expense for other equipment. Because of legal problems with the railroad right-of-way, the canal south of Adams was left intact. North of Blackers was the natural drain.

Raymond Bean offered to fill in the channel at Randalls, but "we were afraid that the heavy equipment would do more damage," Richard said. "Once we made a road so we could get back

Filling in the diversion canal. (Photo courtesy of Nona Jenks.)

and forth with our tractors, the canal did not bother us. We told them we'd take our time and fill it in ourselves, by hand."

From the beginning of the project, church leaders had stressed that the job was only half done when the water was out of the canal. "After the drama is over and the water has dried up, we want to be sure that those who bore the brunt of this aren't going to be out there alone putting up fences and leveling their fields while the rest of the community is out water skiing," Burley West Stake President Walter Petersen told the *South Idaho Press.*

But "the response we had was not as good getting it back as it was tearing it up," he later remembered.

"It's easy to tear things apart, hard to put them back together," said Lane Gochnour. "I didn't have as much help putting things back, but I had some. It was all right. You like to take your time and get it done right."

The farmers were not left completely alone. "Volunteers come out here, pulled out the steel posts and rolled up the wire," said Jack Gochnour. "They cleaned up around the creek. People were really nice. Everybody worked together real good."

Burley Stake President Boyd Poulton organized a work day for his stake. Between 500 and 600 people from ages six to seventy participated. They hauled off sixty to sixty-five loads of rock.

"We had a lot of help," said Richard Randall. "People from Burley Stake came out with shovels and filled in the channel by hand. There were men, women, kids, old people—anyone who could handle a shovel."

"There was all kinds of debris and rocks in my field," said Bart Bowers. "Fifty or a hundred people from Burley Stake came out, walked it, picked up all that junk."

"The whole Burley Stake came out and picked up rocks off my field," said Fred Hawker. "They picked up truckloads of rocks. Scores of people helped pick up rocks. I took a picture of all the cars lined up along the road. Later that fall our ward came with their trucks and spread dirt on my farm where it had eroded."

Wade Baker bought a new blade for his tractor so that he could help push in dirt. "That tractor did a lot of work," said his son Leland. "He was on that thing from clear below the dam all the way through the flood area."

"June 14," wrote Julie Randall. "Wade Baker came again to help. He was here a lot of hours."

Willes Blauer remembered that Wade Baker followed the cat that Ray Bean had leased and leveled off the dirt. "He was out there for days," Willes said.

"Wade Baker and Ray Bean helped us push the canal in," said Marvin Blacker. "Wade Baker spent weeks with his own equipment to see that everyone was helped who could be helped."

To avert a problem in the future, the commissioners arranged for the canal to Murtaugh Lake to remain as a permanent emergency outlet for Goose Creek Reservoir. Leland Baker, Mike Cranney, Roland Bean and others gathered signatures to organize three flood districts.

Next came the task of paying for the canals. The federal government paid for the Corps of Engineers and the equipment they hired, and for the two-weeks' training for the National Guard. The state paid for the Guard's first five days. Still to be paid were the construction costs for the part of the project that Ernie Straubhaar had supervised. Anyone who participated could submit a bill, including those who had volunteered their time and equipment. Some volunteers submitted bills; most did not. "A few people took advantage of the situation," said Leland Baker. "There's always a bad apple somewhere. But for the few who took advantage, there were so many hundreds who were just the opposite."

We knocked some money off our bill," said Paula Watterson. "A lot of other construction companies did the same."

"I did not turn in a bill," said surveyor Steve Pearson. "What I did was so little in comparison. I would have lost so much more if they had not dug the canal. Those who ended up having to sacrifice farmland and things to save us from getting damaged—it still chokes me up to think about it."

Though construction companies from Rexburg had volunteered their time and equipment, they received payment. "We got a check about three months later," said Milo Siepert. "We didn't plan on it. We were shocked when it came."

But they were not paid by Cassia County. "I am finally submitting to you a record of the costs incurred by the Rexburg Region in connection with the work on the Oakley canal project," wrote regional representative Bruce Hafen to Rex Gerratt on July 13. "If it is not possible to obtain reimbursement of the costs through governmental or other sources, we want you to consider this a collective contribution from the people of the Rexburg Region. The people of this area will always appreciate the great support your own people provided them at the time of the flood here in 1976."

County officials talked with the canal board about sharing other costs. "The county would like the Canal Company to be responsible for 1. Raising the spillway, 2. Structural work on the West Canal, 3. Cottonwood canal water," wrote Klyda Wake in the minutes for June 6. "Ennis Pickett stated that we want to be fair. Had not the canals been built we would probably have lost much of our canal system, diversion gates, etc."

The commissioners also hoped to obtain disaster funds from the federal government. The county would have received government money if the reservoir had overflowed. The officials asked for a much smaller amount to defray the costs of prevention.

On July 9 Senator James McClure flew to Burley. He and other officials toured the canal. A meeting was held to discuss funding, then Senator McClure spoke at a Chamber of Commerce luncheon. His office would "try to help determine what can be done to help pay for the project on the federal level," he said, but because the damage was "man-caused," the farmers did not qualify for disaster relief. The federal government supplied funds only for disaster, not for the prevention of disaster.

By August it was clear that there would be no additional outside funds. The county would have to pay the $1.3 million dollars in construction costs. On August 6 the commissioners issued registered warrants—bank loans—for $1.3 million. To redeem the warrants, they authorized a one-time increase in the mill levy for 1985, increasing property taxes by 27 percent.

Everyone had to pay, the farmers whose lands had been damaged and those whose lands had been saved. The people in eastern Cassia County had to pay, including those at Raft River with flooding problems of their own. It was not a fair tax, Weldon Beck told the *South Idaho Press*, nor was it fair that the county was not reimbursing the farmers. But the law clearly stated that they could not pay for damages as the result of a police action, and a property tax was their only means of raising revenue.

A citizens committee, with Dr. Leslie Fillmore as co-chairman, solicited funds to reimburse the landowners. The money came in slowly. At first people hesitated, thinking that the federal government would provide disaster relief. Later, when they learned that their taxes would be raised, many felt that was contribution enough.

"The agricultural economy has been pretty rough, and that has had an effect on the amount people feel they can contribute," co-chairman John Evans, Jr. told the *South Idaho Press.*

"Donations are harder to get after the water stops flowing," said Rex Gerratt.

The committee hoped to raise $200,000. About $40,000 was contributed. "Dr. Fillmore worked so hard," said Rex. "He felt so bad when people were not more responsive."

The commissioners and LDS church leaders had earlier advised landowners to file tort claims that would enable them to sue for damages. Landowners and the railroad filed claims totaling over $1.5 million; the canal company and county filed claims against each other and against the Department of Water Resources for $1 million each.

But though they had filed tort claims, the farmers did not initiate law suits. Many had no plans to sue.

"Who would you sue?" said Jack Gochnour. "It never entered my mind to sue. We're not suing people."

"It's our community," said Richard Randall. "We have to protect the town and community the same as the next person. Somebody has to get hurt to help others."

"It was something that had to be done," said Marian Harris. "You have to be willing to be a part. You have an obligation to society to do the things that are moral and right to do."

While the canal was being built, "a guy running for political office kept encouraging me to file a lawsuit," said Val Robins. "He kept calling and calling. It made me disgusted. He was doing it for himself—wanted to get his name in the papers. I wasn't going to sue. The water had to go somewhere. That's the way we all felt."

Yet Weldon Beck was afraid of lawsuits. He had seen people sue over matters of far less importance. Winslow Whiteley was worried. "We expected law suits in the millions of dollars," he said.

The $40,000 which the citizens committee had raised could not begin to pay for the damages. The farm economy was still depressed. In spite of their reluctance to sue, some farmers were afraid they could not continue in their operations if they did not receive some compensation.

"These landowners will be taken care of," Rex Gerratt had said when the decision was made to build the canal. "I promise you and them, we will not let them down." He knew the landowners trusted him to fulfill that promise.

Once more Rex called a meeting for the three stake presidencies. There would be no outside help, he told them. The stake presidencies needed to take responsibility to see that the landowners were reimbursed.

"Discussion followed," wrote Donald Clark. "The following plan developed:

"Have the Bishops in the channel area personally contact the landowners, member and non-member, discuss their problems as to what would be necessary to correct their damage on a realistic, reasonable actual minimum basis. Servicing of debt and crop loss will have to be included. . . . Compile these figures and set this as a fund raising standard. . . . The Bishops have 2 weeks to assemble their information. The finance drive will start on Sept. 4th."

Farmers filled out a "Potential Loss Worksheet," listing "crop ground displaced by the canal, crop lost due to lack of irrigation, personal assets destroyed."

"People went to the bishop," said Rex Gerratt. "They were asked to figure what they were damaged and then what they would settle for. Their requests for compensation were a fraction of what they had filed for in tort claims. Some did not ask for compensation. People do not realize the sacrifice that was made."

The bishops talked with each member and nonmember alone, remembered Mike Cranney, bishop of the Oakley Third Ward. "We told them to submit for restitution what they felt good about. Most people wrote off their expenses. Most people on Goose Creek did not take restitution. They figured they would have been damaged anyway. Our people were super good about it."

Dick Whiting, a landowner who had received minimal damage, turned in the following worksheet:

POTENTIAL LOSS WORKSHEET

CROP GROUND ~~DISPLACED~~ *replaced* **BY CANAL:**

We gained 3 feet of topsoil in our garden but we didn't order it and we are not going to pay for it.

CROP LOST DUE TO ~~LACK OF~~ *over* **IRRIGATION CAUSED BY CANAL:**

2 rows carrots, 4 rows peas, 2 rows beans, 1 row squash, 1 row pumpkin, 1 row lettuce, 2 rows corn, 1 row tomatoes, 1 rhubarb plant.

PERSONAL ASSETS DESTROYED BY CANAL:

I can't find my fence but it's got to be under there somewhere.

OTHER ITEMS CONSIDERED A LOSS AS A RESULT OF CANAL (Describe each item fully and list value):

Hair, sleep. Great personal worth, very little commercial value.

At the bottom of the worksheet he wrote, "Anything we may have lost is more than compensated for in the knowledge that we live among men of vision and that we have neighbors that will be there if we ever really need them."

"I asked for money to pay for the crops," said Fred Hawker. "We figured the damage was on us. I tried to be fair and honest. The canal had come through nine acres. My bishop said, 'You haven't lost that much, have you?' I thought, 'my own bishop, didn't even know.' I told him to come and look. After he'd looked at it then he could see it. The problem was you couldn't see any damage from the road."

"It was proper that the Bishop should be responsible for satisfying these claims," wrote Donald Clark, "since the judgment should be between the landowner and the Lord and the Bishop to assist."

After hearing from the bishops, the stake presidents set $300,000 as a fund-raising goal. They asked the city of Burley and the business community for $100,000 and the canal company for $100,000. The three stakes would raise the additional $100,000.

"Rex Gerratt met with the board to discuss the settlement of claims filed against the County, Canal Company and Dept. of Water Resources," wrote Klyda Wake in the September 17 canal company minutes. "It is his feeling that these can be settled fairly and peaceably without prolonged litigation and excessive expense. He expected to know by Sunday a total cost for settlement."

On September 29 she wrote: "Time was turned to Mr. Gerratt who reported on meeting with area farmers along the Oakley to Burley canal and figured that they would settle for around $300,000. . . . Mr. Gerratt feels it isn't unreasonable to ask the Canal Co. to stand some of this cost of reimbursement due to work done on existing west canal. . . . Motion was made by Jay L. Robinson that the company submit $100,000 now to President Gerratt's committee. . . . Voting was unanimous by Board."

Grant Wyatt and Cloyd Taylor were assigned to raise money from the city and businesses. They asked the city of Burley for $50,000. The city did not contribute. "We were informed by our attorney that according to the Idaho code the city could not participate in a county disaster," said Mayor Chuck Shadduck. "We felt bad about it, but there was nothing we could do."

Grant and Cloyd next approached the businesses. "I'd had experience in asking for money for the Boy Scouts," said Cloyd. "If you go to somebody's door and ask for money, you get fifty cents or a dollar, whatever they have in their pockets. So we'd call people on the telephone, tell them it takes $40 to sponsor a boy in Scouting. Would they be willing to sponsor a boy, or a half a boy? That way we'd get some real money.

"When we approached the businesses, we asked them to think about what it was worth not to have water on their place, to think about the millions of dollars they would have lost if the water had come. We asked them to weigh that with what they might give. What we asked for was so minute compared to the potential damage that there was no comparison."

Though a few said no, "most came through with sizeable contributions," Cloyd said. "Some county people felt that Burley did not pick up its fair share, but people within the city contributed a substantial amount."

The three stake presidents and the bishops worked with their stake and ward members to raise the final $100,000. A certified public accountant audited all the funds that were collected.

"Many people were the beneficiaries," said President Walter Petersen. "It was appropriate to compensate those who sacrificed. All wards came up with the money quickly. It was gratifying to see the financial contributions."

The Burley Fourth Ward sent letters to their members. "Last spring when many of our homes and properties were threatened by flood water, a few people willingly sacrificed their property to save the rest of us from tragedy," the letter said. "And if you saw the size of the canal, and the amount of water it carried, you must realize that without the canal we truly would have experienced a tragedy. . . .

"Our request is that each family in the Fourth Ward contribute at least $[xxxx] to the flood fund.

"We are painfully aware of the difficult financial circumstances that many people are in. We are also aware, however, that each of us would have been in a much worse financial circumstance had the flood waters actually come our way. . . . "We humbly make this request for contributions, and in confidence promise you the blessings of the Lord if you will so sacrifice."

In some wards bishops made appointments with members to ask for contributions. Some persons who had contributed to the civic fund made additional donations. One elderly woman, who had donated to the civic fund, wanted to give more, but did not feel she could afford to. Her son who was living in another state made a contribution in her behalf.

They had no trouble raising money in their ward, Denny Davis remembered, even though the ward was located in an older, less affluent section of town. "Those good, faithful older people—the ones who had always been there helping out over the years—they didn't have a lot, but they gave what they had," he said. "They knew what they had been saved from, and they were grateful."

"November 26," wrote Donald Clark. "President Gerratt reported on a visit with Elder Zachreson, our Regional Representative, in regards to handling of moneys collected and distribution to the landowners. He gave a vote of confidence as to the progress thus far and agreed that the funds be dispersed as soon as possible."

On December 11 President Gerratt and President Petersen met at the Burley Stake Center in separate rooms. Each president met privately with each landowner in his stake who had flood damage, including the landowners who were not members of the LDS church.

"I'd sit across the desk from someone," said President Gerratt. "We'd talk about the emotions, about the physical damage. Then I'd say, 'I know you've talked to your Heavenly Father about this. What would it take to make you feel good?' They'd tell me, and I'd make out a check. The settlement was between the landowner and the Lord."

The amount each received was confidential. The landowners signed waivers stating that they would not seek litigation. Several landowners would not accept any reimbursement. They also signed waivers.

A short time later the $40,000 which the citizen's committee had collected was divided among landowners according to the linear feet of canal that had come through their property. The Episcopal Church of St. James and some other organizations had held fund raisers earlier and made donations to landowners. Glen Jensen, ASCS administrator, provided some additional compensation. "The government cooperated any way they could with their annual programs," he said.

Even with this money, the landowners were not totally reimbursed for what they had lost. "The compensation was not equal to the damage," said Cloyd Taylor. "How do you compensate when a person loses topsoil? A few sacrificed to save thousands."

"There's not enough money in the county to pay for all the damage that was done," said Marvin Blacker. "People are just living. They helped as much as they could—or thought they could. A lot of people contributed money that no one would ever know about."

"No one was fully compensated," said President Gerratt. For each landowner it had been, and would remain, a sacrifice.

But others had also sacrificed. Some farmers had furnished gas and equipment and paid their hired help to work on the canal. They neglected their own farms. Some people worked eighteen-to-twenty-hour days on the canals for a month or more. A volunteer who kept track of his mileage found that he had driven 1,800 miles delivering lunches. Professional people left their offices and businesses. Some wage earners took time off work without pay. "To this day I don't know how our family survived without an income for two weeks," one volunteer wrote.

"I think of all the people—those in Unity and View—who wouldn't have been affected, how they came over and helped," said Val Robins. "Somebody in town donated a lot of money for it."

Donald Clark, who tried to keep track of the volunteer labor that the LDS church recruited, estimated that approximately 2,600 volunteers contributed 40,000 labor hours and 10,000 equipment hours. "These are conservative figures," he wrote. "Some of the heaviest contributors refused to turn in their hours. They simply do not want this information known or printed."

"Some people had $100,000 apiece in the project and did not put in a claim," said Rex Gerratt.

"Lots of people donated a lot of time without compensation," said Leland Baker. "I could have turned in a big bill, but I didn't feel right about trying to collect anything on that flood deal. It's something you do. It's something you owe your country, county, state, family, and friends. It's payment for the space you take up on this earth."

None of the landowners sued. The county and canal company did not sue each other. "There was so much potential for legal disputes," said one lawyer. "Then they all just evaporated."

"It amazes me that so much could have been done so fast with so little opposition," said attorney Steve Bywater. "I have seen problems much less significant than this one dragged through the courts forever. There was a willingness of the individual to sacrifice for the common good."

"Salmon Falls dam flooded at the same time," attorney Al Barrus said. "There was a volume of litigation on that. It went to a jury trial. The litigation is still going on. We had no lawsuits because we had wonderful people involved."

Union Pacific Railroad officials, however, did bring suit against the county. They lost their suit. The county's legal defense showed that a bona fide emergency had existed in Cassia County and that the county had complied with the law when it appropriated railroad property.

The canal company and railroad then negotiated an out-of-court settlement in which the canal company paid for the damaged railroad track and purchased the railroad right-of-way. The company sold the right-of-way to farmers whose lands were adjacent to the railroad for the price the company had paid for it.

But even as legal and financial problems were being settled, a new crisis arose.

Throughout the emergency, people had worried about the reservoir gates, but no one had thought about the tunnel. From the gate opening the tunnel ran underneath the reservoir floor, past the tower, and under the dam. At the end of the irrigation season, Jay Gorringe noticed that rocks, pieces of concrete, and other debris had been carried from the tunnel into the outlet canal. The Department of Water Resources sent Allen Merritt to inspect the tunnel.

"Jay had a shovel and I had a survey rod," Allen said. "All the way back to where the tower is, was pretty clean. After we went past the tower, we could hear rushing water. We worked our way around one hole. Then Jay fell in a hole—just disappeared. We got back out and called the state office. They got float tubes and made it back all the way.

"The holes in the tunnel were unbelievable," Allen said. "The water had eaten into the concrete liner. All the concrete below the gates was stripped out, the plate steel rolled off and broken up with the force of the water."

Between the tower and the gates the tunnel made a ninety degree turn. The water hit the wall, bounced to the ceiling, and eroded the basaltic lava rock. About fifteen feet separated the top of the hole from the floor of the reservoir.

The canal company asked Dal Wayment to assess the damage. "It was three days before Christmas," Dal said. "The Department of Water Resources was concerned. It was one of the scariest things I've done in my life—to go back in the tunnel with those guys. The leakage put a lot of water in there. We had to float in on tubes. It was noisy and cold—a long, dark tunnel.

"The tunnel is perpendicular from the gates into the face of the dam, then it goes at an angle. The water created a boil there. The lava rock was pretty fractured. The water was coming into the hole from the bottom of the dam at a tremendous rate. Because of the tower and the water pouring out, there was a ten- to fifteen-mile breeze blowing from the tunnel. Water was falling a hundred feet from the tower and shooting into the tunnel. We had to shout to communicate. Streams of water were shooting in from the side."

No one knew exactly how much rock was between the hole and the reservoir floor, how strong the rock was, nor how long it would hold. But if the tunnel failed, it would drain the reservoir. Once more Terry Bingham, canal company directors, commissioners, and church leaders huddled together in emergency meetings. Once more those in the water's path were told to prepare. Once more the Corps of Engineers was asked to help.

Dal met with some tunneling companies in Salt Lake City, then with Corps emergency personnel and the Department of Water Resources in Boise. "I suggested that we build a bulkhead around the hole and pump special grout into the fissures," he said. "In a couple of days the Corps put together directions for a contractor. Within seven days the contract was let out. They moved in and drained the tunnel. It worked until they could drain the reservoir."

The next year the canal company lined the tunnel with steel pipe with six-inch-thick walls, and encased it in concrete. "We bonded for $1 million to pay for it, plus some other moneys we had on hand," said Winslow Whiteley.

The crisis was finally over. The Soil Conservation Service cited the Burley office for its work. "We received a plaque to hang in our office," said J.J. McLaws. "Everyone who participated received a cash award."

Each National Guardsman who helped with the canal was awarded a medal and a ribbon.

Engineering News Record nominated Terry Bingham and Tom Tate for Construction's Man of the Year award. February 13, 1985, they flew to New York, where they were among forty-nine nominees who received recognition for their outstanding construction work.

The crisis was over, but other evidence of destruction remained. The canal had changed the land—dropping silt in some places, eroding topsoil in others.

The channel through Harrises' property was twenty feet deep and a quarter mile long. "It would cost ten times as much to replace it as what the quarter acre was worth," Ed Harris said. With their reimbursement money and help from the SCS they filled in as much as they could with tare dirt which Ore-Ida gave them. "We're still trying to square the field out and push dirt around," said Ed. "I still get dirt from Ore-Ida on occasion to help fill in a hole."

Jack Gochnour used wire and baled straw to build checks every hundred feet along the channel through his land. The checks catch the silt as it washes down, gradually filling in the channel. "I can farm a little more of it every year," he said.

After losing some of their topsoil a second time during a storm, the Hawkers buried bales of straw to keep their fill dirt from washing. "We keep building it up," said Fred. "One good thing—I love to move dirt."

Other, less tangible effects were part of the aftermath. It had been a traumatic experience for all who had been heavily involved, "a numbing, shocking deal," as Lane Gochnour called it.

"It was like having an accident victim in your family, like losing a loved one," said Lovel Blacker. "The shock, the anger, the mourning—all the steps that you go through with a death—they were all there. You had to overcome, had to repair your life."

"I don't ever talk about it," said Marvin Blacker. "I guess I haven't figured out yet how to handle it. I remember I wrote one sentence about it in my journal. But I want the generations to know."

Some people had said things thoughtlessly; some said things in anger. There were feelings, hurts, and resentments that could not be resolved quickly or easily.

Yet even as they covered the canal and planted their crops, people began to put the emotional trauma behind them.

"I had some bitter feelings I had to overcome," said Lane Gochnour. "I grew from it—had to grow or die. I feel I'm a bigger man now because of it, a better man."

"Time is a healer," said Marian Harris. "Time softens all things."

Winslow Whiteley remembered that Marvin Blacker came to him and others and apologized for anything he might have said.

"The farmers thought, 'I'm required to pay an awful price for someone else's problem,'" said Rex Gerratt. "The Lord compensates. People have been blessed."

A few years after the flood, Willes Blauer moved to Burley—to a house in the old Goose Creek channel. He was called to be bishop of the ward there. "You have a lot of different views," he said as he reflected on his experiences. "Knowing these people now—wonderful people—to have them flooded would have been a disaster."

"Hindsight is better than foresight," said Val Robins. "We know now that it was the best way to do it."

"Life gets humdrum at times," Fred Hawker reflected. "Something like this pulls people close together. If nothing ever happens, you never find out what kind of neighborhood you live in. This is the best neighborhood in the world. We're in a community where people help people."

Today no one talks about the canals. Everyone is involved in concerns of the present as they hope for high commodity prices and pray for rain. On the surface life goes on as if the crisis had never happened.

But only on the surface. For the building of the canals was more than just the saving of a community: it touched something deep and fundamental in people's lives. The experience of the canals became for people an affirmation of faith—a tangible witness of the goodness of the individual, an evidence of the blessings of the Lord.

Some pointed to the weather as evidence of the Lord's care. "Two cold nights hit exactly the right time," said Mike Cranney.

"The good Lord gave us about two days to get some things done," said Dallan Elquist.

Many mentioned the lack of serious injury. "It was a miracle no one was killed," said Bill Cranney, "with all that heavy equipment working day and night. There were several mishaps, several close calls."

"Nobody was sleeping right or eating right," said Kim Cranney. "They were hauling rock at night. Inexperienced people were operating dump trucks—the most dangerous things there are."

"We were going twenty-four hours a day, working in the dark, a lot of equipment had no lights," Leland Baker said. "There were no serious injuries or deaths."

Others pointed to the timing of the events. If any one thing had not been completed on time, the whole project would have collapsed. If the canal to Murtaugh Lake had not been in place when Cottonwood overflowed, that creek would have flooded where the second canal was later built. If the spillway had not been raised, there would have been no time for the second canal. For decades field stones had been stacked in piles in the desert, ready to use when they were needed. The few feet of rock separating the tunnel from the reservoir held until the tunnel could be repaired.

"If we had known ahead of time the amount of water that there would be, we could not have made better decisions," said Roland Bean. "Everything was done the way it should have been. People were inspired all the way through. I remember a feeling I had while on the road, 'This is not going to be easy, but it can't fail.' I knew we could do it. There was no doubt in my mind."

"There was tremendous cooperation," said Grant Wyatt. "People were eager to help. It restores your faith in people."

So many people and groups were needed—the Corps of Engineers, Soil Conservation Service, Department of Water Resources, Bureau of Reclamation, Bureau of Land Management, National Guard, Civil Defense, County Commission, Oakley Canal Company, highway and street departments, police and sheriff's departments, Burley city services departments, Twin Falls and Milner canal companies, Kloepfer's and other construction companies, farmers, volunteers, landowners, the Salvation Army, the LDS church, other churches—all played significant roles. The success of the project depended on all. Without the cooperation and contributions made by each group and individual, the project would have failed.

"There is no way you could get everybody's names who contributed," said Burley Police Lieutenant Jim Higens. "Some people did truly heroic things—others did the little things. But everyone's contribution, no matter how small, was extremely important. There were a lot of unsung heroes."

The construction companies did not sign contracts, Earl Corless remembered. "People just went to work without knowing they'd get paid. I don't know how many places in the world you could get something like that done."

"There was real wisdom in the project," said Kent Searle. "You can't credit all one person. Terry did an excellent job. Rex had the leadership that pulled people together. The Army Corps people were good to work with, the National Guard, Ernie Straubhaar, the youth with their sandbags, the lunches, all the people responding to whatever needed to be done.

"For the farmers it was a real sacrifice. When it's your own ground, you've bought it and worked it, it becomes part of you. If they had not been willing, we could have been tied up in litigation forever.

"There are politicians and there are statesmen. We had statesmen in this county. They were concerned about the people."

"We were a blessed people," said Burley Stake President Boyd Poulton. "Everyone rallied around and worked together. We did not use police power at any time. We were able to resolve any difficulties without police power."

Today those who were involved in the project cannot talk of it without becoming a little emotional.

"It was a highlight of my life," said Burley West Stake President Walter Petersen, "to see the willingness of the people to do what needs to be done when it needs to be done. In a real emergency people rise to the occasion. It was gratifying to see the cooperation and unselfishness. It made us realize that our Heavenly Father has his hand in what happens on this earth."

"The credit goes to the wonderful people," said President Gerratt. "They didn't wait for a government declaration of emergency or red tape, everybody did their share. I never want to go through it again, but we have enough resources. The strength for doing what needs to be done lies within the people. Good, solid, commonsense, hardworking people made it run. The route and the decisions along the route were made known by inspiration and were answers to a lot of humble prayers. It was a test. People don't understand what everybody went through. It was a controversial thing, but they were good enough that they did it."

"It was a once-in-a-lifetime experience," said Terry Bingham. "Normally in my job as Deputy Sheriff I see the dark side of people. That project gave me a confirmation of the positive side of the human spirit that is enough to last a lifetime."

Those involved would not take credit for the accomplishment.

"I was just doing my job," said Tom Tate.

"I was in the right position at the right time," said Terry Bingham. "I didn't move one shovelful of earth. I'm not a hero, but I was in the midst of heroes."

"The Hero was a lot higher than me," said Leland Baker. "I remember Bill Cosby saying something like 'There's a lot of people that think they're atheists, but if the ground shakes, you're going to look up.' That's what happened to me. The ground shook, and I looked up."

"I would like people to know that the Lord directed our actions," said President Gerratt. "That's one of the reasons why it went so very smooth. Dozens and dozens of things I said that were out of character for me. Those directions just came in time. You knew the Lord was directing the project. We were being touched to know what the best judgment was. Those feelings I can't describe. We were calling on the Lord and needing his help. It came because people were willing to cooperate and follow their leaders."

"An example of present-day inspiration is the Oakley flood diversion project," said Lane Gochnour in a talk he gave a year later in church. "The Oakley flood diversion project was inspired. None of the church leaders and I don't think any of the civic leaders involved in the Oakley project would take credit for this miracle even if you twisted their arm because anyone involved closely with this project knows that it was the spirit of the Lord that enabled a handful of property owners and an army of willing people to cooperate, put their personal considerations aside and help God to accomplish one of the greatest miracles of this generation."

Lower Goose Creek Reservoir, July, 1992.
(Photo courtesy of Jeffrey Hedberg.)

Epilogue, 1992

As you drive along the Oakley-Burley highway today, it is difficult to imagine that a flood once threatened this beautiful valley. The years since 1985 have been dry. Looking at the small amount of water in the bottom of Lower Goose Creek Reservoir, it is hard to realize that once it was full, ready to run over.

The boards which held back the flood waters are no longer in place, but the one-foot cement addition and channel irons are still on top of the spillway. The canal to Murtaugh is filled with dried brush and weeds. Since 1985 it has carried no water. Only a few remnants of the Snake River diversion canal remain, though the farmers can show you where it crossed their fields.

Some of the persons who participated in the events of 1984 are no longer here. Kent Foster, Burke Scholer, and Evan Rasmussen now live in Boise; Harold West is in Arizona; Leo Busch is in Denver. Jay Gorringe is deceased, as are Wade Baker, Grant Randall, Leon Adams, Harold Bowers, and Dorothy Lattimer. Klyda Wake and J.J. McLaws have retired. Although Rex Gerratt still works on his dairy farm with his family, he is no longer the Oakley Stake President, but serves his church as a regional representative.

Some things have not changed. Ennis Pickett and Daren Critchfield are still members of the canal board, with Winslow Whiteley, now eighty-two years old, still serving as president. Terry Bingham is civil defense director, and Weldon Beck, who decided that he would run for office again after all, is at age eighty-two still a county commissioner.

In Oakley's park close to the center of town, you will find a monument made of Oakley stone which contains a bronze plaque with this inscription:

IN TRIBUTE TO THE FORESIGHT AND WILLING EXPENDITURE OF TIME, EQUIPMENT, TALENT, AND RESOURCES BY MEN, WOMEN, GOVERNMENT AGENCIES AND CONTRACTORS FROM THIS AND OTHER AREAS GUIDED BY THE HAND OF GOD AS THEY WORKED SHOULDER TO SHOULDER IN A RACE WITH TIME TO DIVERT AN EMINENT DISASTER TO LIVES AND HOME AND PROPERTY IN DIVERTING THE OVERFLOW FLOOD WATER FROM THE OAKLEY RESERVOIR SAFELY TO MURTAUGH LAKE AND THE SNAKE RIVER IN THE SPRING OF 1984.

Lower Goose Creek Reservoir, spring 1984.

(Photo courtesy of Nona Jenks.)

Appendix A

(Remarks made by Oakley Stake President D. Rex Gerratt at the Oakley Flood Control Appreciation Fireside held June 28, 1984, at the Burley Stake Center, as reconstructed by President Gerratt from minutes taken by Historical Clerk Donald J. Clark.)

Dear brothers and sisters, neighbors and friends. With great appreciation we welcome you here tonight. We give a special welcome to Elder Hugh W. Pinnock, of the First Quorum of the Seventy, and Elder Martin K. Zachreson, Regional Representative, and thank them for their constant support.

I have never had anything touch my heart as much as the experience of the last two months in our county. I want to express my love and appreciation for the spirit, support, and attitudes of all who were involved in any way in building the canal. To everyone, whether members of the LDS Church or not, I am so grateful to be able to call you my friends.

It is a great feeling to work with people who don't know that things can't be done. Everything that was done had to fit within an exact time frame. Only a very few hours made the difference in the success of the whole operation. There was no time for long, controversial debate.

The role of the LDS Church has been to work with and support all civic leaders that had charge of this project. This project was, in fact, inspired by our Heavenly Father. I testify to that. It was guided and prompted daily by His inspiration. Verification came on numerous occasions as the project unfolded. It rates among the greatest miracles that I have ever read about or witnessed. All credit should be given to our living, loving Heavenly Father. Regardless of what church you belong to, you should get down on your knees each and every day and thank your Heavenly Father for what has been accomplished.

The miracle was not that an 18-mile canal was built in four days to Murtaugh, and that a 24-mile canal was built in three days from Oakley to Burley, because there is no limit to the amount of equipment that could be brought in for such a project. The miracle was in the attitudes and willingness of the people to give, cooperate and do whatever they were asked. As people were called, I stood in tears, watching them come to a job site in only a few minutes, willing to do whatever was necessary, even in the middle of the night.

Another miracle is the way irrigation pipes were moved, telephone lines relocated, surveying done, easements obtained, equipment located and enlisted—each decision and its completion was a miracle. Government officials estimate that just to do the preliminary work for the Oakley Canal alone would have taken two-and-a-half years under ordinary circumstances.

To those whose land the canal crossed, we thank you from the bottom of our hearts. You now own some of the most valuable land anywhere in the world—thinking of its worth in the damage saved. Those of you who own this land, if your attitude is right, that strip of land that runs through your farm will be one of the greatest sources of blessings you will ever have in your lives. You will not want to part with that strip of land, because it is a monument to true love shown and much sacrifice given.

We express our gratitude to people in other areas for their concern and assistance. The stake presidencies have become very close. We have been on our knees every night together for several weeks, as we have talked about the days, and prayed about the tomorrows, as we have worked with those who had responsibility.

We are all recipients of a blessing asked by Elder David B. Haight as he dedicated the Grouse Creek chapel in the Oakley Stake on May sixth. He petitioned the Lord, "that the weather might be steady, and the snow will not melt too fast, and that the water will be able to be taken care of, and the people be spared from serious flooding."

Brothers and sisters, I testify to you that the elements were tempered, that the inflow into the reservoir did slow, just enough that we were able to get approval to build a three-foot extension to the spillway. That extension gave us the 72 hours necessary to complete the building of the canal from Oakley to Burley. When the water was turned in the canal, at 8 A.M. Sunday morning, May twentieth, there were six miles of channel yet to finish on the north. It was finished in time for the water to run in the Snake River at 8:15 A.M. the next morning.

This experience has done much in my life to renew my faith in people, in this community, and in America. Government assistance was put in proper perspective. The government agencies worked for the people. There were no run-aways or empires built. Everyone worked together.

Brothers and sisters, we are not finished yet. The bills are not paid. We need to tell our neighbors, our friends, and business associates. We need everyone to help see this project through to a successful conclusion. As stake presidents we have so committed and expect all of you to support us.

In closing, may I again express my appreciation to all who helped in any way. In the name of Jesus Christ, Amen.

Appendix B

How much water was there? In the spring of 1984 the U.S. Geological Service gauge on Goose Creek malfunctioned during peak inflow. Later, USGS officials estimated the amount of water that flowed into the reservoir. However, a comparison of the estimated figures with the figures for reservoir increase shows that the estimates for water flow were too low (see Figures B–1 and B–2).

Goose Creek gauge is located five miles upstream from the dam. A mercury manometer gauge, it continuously records the stream height. Officials use a chart to convert stream height into cfs. Every two months USGS officials measure the flow and adjust the chart as stream conditions change. During a flood they measure more frequently.

In May 1984 USGS hydrologist technician Mike Campbell measured Goose Creek May 10 and May 15. At 10:23 A.M. Tuesday, May 15, Mike began measuring the flow at Goose Creek, taking about an hour to complete his measurement. At that time the Goose Creek gauge was functioning. Later, someone removed the gauge cover, and when replacing it, pinched the gas line. This caused the gauge to give a false reading.

Officials were not aware of the problem. Corps of Engineers hydrologist Bob Rickel, who forecast streamflow based on the Goose Creek gauge, concluded that the peak had passed. But when the discrepancy between the gauge reading and the increase in reservoir height became apparent, Bob asked the USGS officials to check their gauge. Joe Bunt checked the gauge Thursday, May 17, at 6:15 P.M., discovered the pinched line, repaired it, and made a stream measurement. Thus the line was pinched for about fifty hours, from sometime after 11:30 A.M. May 15, until 6:15 P.M. May 17. As a result there is no accurate data to show how much water entered the reservoir during that time.

After the crisis was over, USGS officials estimated the amount of runoff during those fifty hours. Historically, Trapper Creek and Salmon Falls Creek had flow patterns similar to Goose Creek; therefore, officials estimated the Goose Creek streamflow by using data from Trapper and Salmon Falls creeks. Because those creeks had peaked May 17, the officials concluded that Goose Creek also peaked that day. Goose Creek was flowing at 1,240 cfs when Mike Campbell measured it at 10:30 A.M. May 15; the Goose Creek gauge recorded 1,320 cfs as the average flow for May 15, with 1,590 cfs as the peak flow. The officials used the 1,320 cfs figure

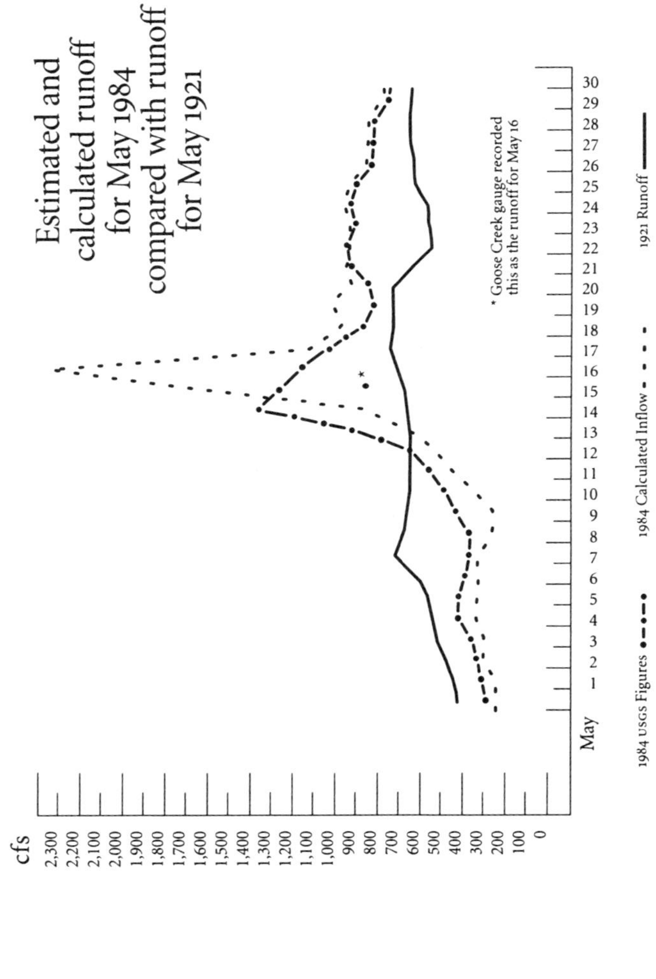
Estimated and calculated runoff for May 1984 compared with runoff for May 1921
cfs
2,300
2,200
2,100
2,000
1,900
1,800
1,700
1,600
1,500
1,400
1,300
1,200
1,100
1,000
900
800
700
600
500
400
300
200
100
0
May
1
2
3
4
5
6
7
8
9
10
11
12
13
14
15
16
17
18
19
20
21
22
23
24
25
26
27
28
29
30
* Goose Creek gauge recorded this as the runoff for May 16
1984 USGS Figures
1984 Calculated Inflow
1921 Runoff

Figure B–1

Figure B–2

1	2	3	4	5	6	7	8	9	10
April 30	52 36	0.10	120.92	58,200	58,220	—	—	—	—
May 1	52 36	0.06	121.12		58,338	99	145	244	277
2	54 38	0.12	121.32	58,600	58,534	99		244	308
3	55 37		121.65	58,900	58,857	164		309	335
4	52 32		121.98	59,300	59,082	163	134	297	373
5	53 36	0.08	122.42	59,700	59,616	218		352	411
Sunday 6	50 27		122.82		60,012	199		333	410
7	59 28		123.22	60,500	60,406	199		333	392
8	60 32		123.61	60,900	60,788	197	130	327	370
9	68 36		123.82	61,100	60,994	109	170	279	386
10	74 38		124.02	61,300	61,191	104	172	276	437
11	65 43		124.42	61,700	61,603	208	173	381	495
12	70 38		125.02	62,300	62,220	302	160	462	570
Sunday 13	86 48		125.72	63,000	62,920	353	160	513	666
14	84 53	T.	126.67	64,000	63,870	479	135	614	908
15	70 35	0.30	128.12	65,400	63,321	749		884	1,470
16	52 34	0.09	130.57	69,100	67,977	1,351	135	1,486	1,282
17	65 32		133.79	71,800	71,646	1,933	311	2,244	1,160
18	69 39		135.05	73,200	73,163	805	320	1,125	1,023
19	75 44		135.93	74,200	74,263	640	347	987	872
Sunday 20	75 52		136.57	75,100	75,063	452	564	1,016	820
21	60 40		136.90	75,500	75,475	255	635*	890	861
22	70 37		136.95	75,600	75,538	37	897**	934	926
23	77 52		136.83	75,400	75,388	-86	1,000	914	927
24	67 46		136.72	75,200	75,250	-78	1,000	922	915
25	63 37		136.63	75,100	75,138	-64	1,049	985	935
26	67 51		136.47	75,000	74,938	-113		936	907
Sunday 27	69 49		136.22	74,600	74,850	-176		873	849
28	80 40		135.96	74,400	74,300	-184	1,049	865	841
29	88 50		135.68	74,000	73,950	-198		851	805
30	84 58		135.33	73,500	74,763	-252		797	786

*USGS measured the outflow as 705 **USGS measured the outflow as 957

Explanation of columns:

1. Date.
2. Maximum and minimum temperature (degrees Fahrenheit) in Oakley.
3. Precipitation (inches).
4. Gauge height or elevation in feet from a given point on the reservoir floor. (These have been recalculated using the USGS base elevation which is .58 feet lower than the Bureau of Reclamation base elevation.)
5. Reservoir storage in acre feet, USGS figures rounded to the nearest hundred.
6. Reservoir storage in acre feet, Bureau of Reclamation figures.
7. Change in reservoir storage, cfs.
8. Outflow in cfs, measured by the Idaho Department of Water Resources. (Where no figure is given, the outflow is assumed to be the same as for the preceding day.)
9. Computed inflow in cfs (change in reservoir storage plus outflow).
10. Measured inflow from Goose and Trapper creeks, USGS figures. (Figures for May 16 and 17 are estimated.)

as the flow for May 15, then estimated flows of 1,140 cfs for May 16, and 1,030 cfs for May 17.

However, these estimated figures do not correlate well with the reservoir storage data that shows the greatest increase on May 17.

Theoretically, adding outflow to the increase in reservoir storage should equal the amount of water entering the reservoir. However, there are problems with this. Reservoir storage is determined by using a 1932 chart to convert reservoir gauge height to acre feet. Although the chart is accurate for lower reservoir levels, it is not known how accurate it is for higher levels. As the reservoir fills, the water spreads into areas that may not have been calculated in the chart. In addition "bank storage" occurs as the water flows into cracks and crevices and soaks into the soil.

In spite of these problems, as Figure B–1 shows, there is a correlation between the inflow as calculated from the increase in reservoir storage and inflow as measured by the Goose and Trapper creek gauges. Between May 1 and May 13 the calculated inflow was 50–100 cfs less than the measured inflow from Goose and Trapper creeks. Calculated inflow was about 300 cfs less on May 14, 600 cfs less on May 15. This likely occurred because of increasing bank storage as the reservoir water spread over a wider surface.

The reservoir reached its highest level May 22 (though this was not the day of its greatest increase). After this date, bank storage would not be a factor because the water would already have soaked into and filled up the cracks and crevices. From May 22 through May 30, the calculated inflow and measured inflow are almost the same. The big discrepancy in the data occurs on May 16 and 17, the days during which the Goose Creek gauge malfunctioned and the inflow was estimated. The correlation between the measured and calculated inflow before and after those two days shows that, if bank storage is taken into account, the calculated inflow is fairly accurate. Thus the calculated inflow is probably more accurate than the estimated inflow which was based on other creeks.

The calculated data also shows that, when bank storage is taken into account, the peak inflow occurred May 16, a day later than the estimated figures indicate. Evidence that the peak did occur on May 16 comes from Goose Creek rancher Morris Mitchell, who lived about eight miles upstream from the dam. On May 16 he wrote in his journal: "Goose Creek [at the ranch] peaked early this morning and started to lower slowly."

Therefore, it appears that about 2,500 more acre feet of water entered the reservoir than was officially estimated, and that the peak inflow occurred Wednesday, May 16.

Anticipated Inflow Compared with Actual Inflow

The streamflow for May 1984 was much greater than had been predicted. Figure B–3 shows Dal Wayment's estimated runoff compared with the actual runoff. Dal used information from the Soil Conservation Service in making this estimate.

Figure B–3

	Projected Runoff	Actual Runoff
March	6,000	6,847
April	16,000	12,639
May	15,500	47,111*
June	10,000	23,106
July	7,000	7,406
Total	54,500	97,109

*Using figures calculated from reservoir storage for May 16 and 17.

Water Delivered to Shareholders

Figure B–4

Year	Inches	Year	Inches	Year	Inches	Year	Inches
1913	18+	1933	10.5	1953	18	1973	24
1914	13+	1934	4.5	1954	15	1974	28.75
1915	12.75	1935	8	1955	9	1975	38
1916	17	1936	13	1956	16.25	1976	24.75
1917	18	1937	11	1957	16.5	1977	18
1918	13.25	1938	13	1958	18	1978	18
1919	9.6	1939	12	1959	10.6	1979	18
1920	8.6	1940	7.25	1960	8.25	1980	18
1921	28	1941	8.5	1961	6.75	1981	18
1922	22	1942	18	1962	18	1982	18
1923	18	1943	18	1963	12.5	1983	18
1924	14.5	1944	18	1964	13	1984	18+
1925	13.8	1945	18	1965	18	1985	18
1926	9.15	1946	18	1966	11.5	1986	18
1927	13.3	1947	13.5	1967	11	1987	10
1928	14.5	1948	11.5	1968	8	1988	10
1929	15.1	1949	18	1969	15	1989	13.5
1930	8.5	1950	18	1970	18	1990	8
1931	6	1951	18	1971	21.25	1991	8
1932	13	1952	18	1972	30	1992	4.5

Figure B–4 shows the annual number of acre inches per share that the Oakley Canal Company delivered to shareholders. Figures are taken from canal company records.

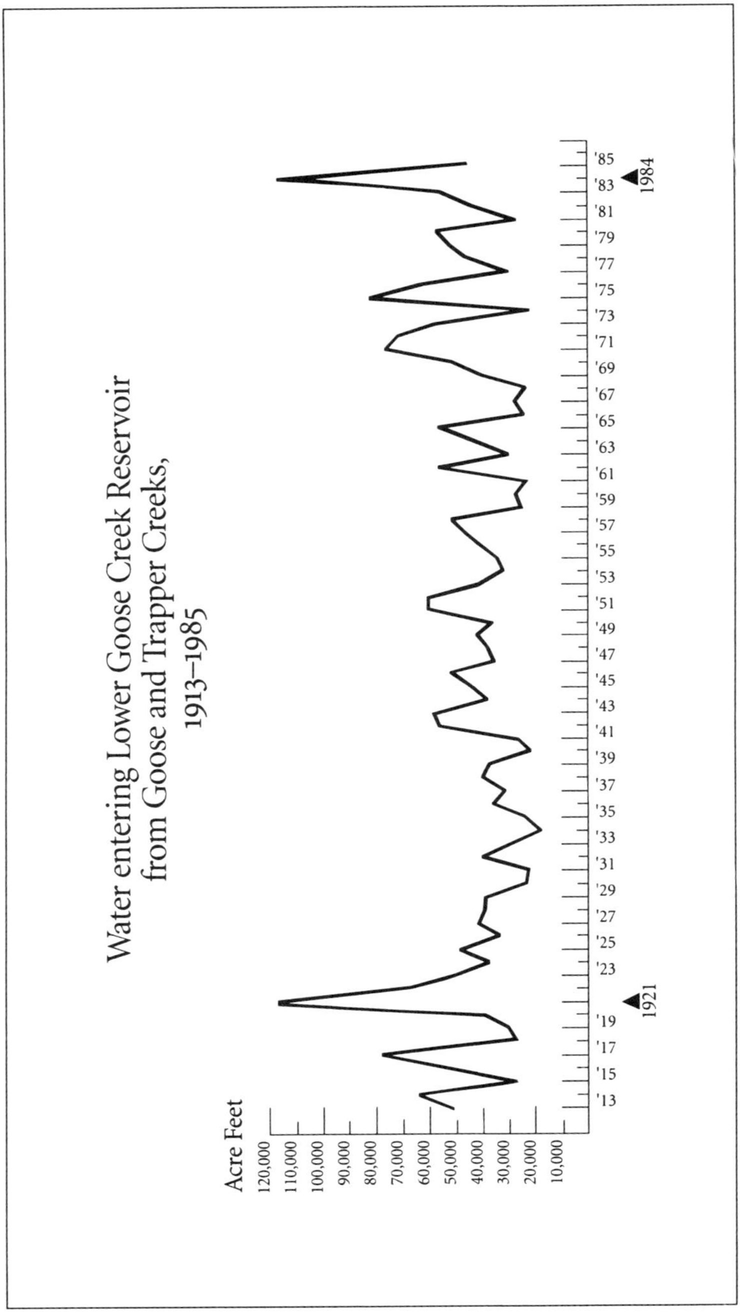
Water entering Lower Goose Creek Reservoir
from Goose and Trapper Creeks,
1913–1985
Acre Feet
120,000
110,000
100,000
90,000
80,000
70,000
60,000
50,000
40,000
30,000
20,000
10,000
'13
'15
'17
'19
1921
'23
'25
'27
'29
'31
'33
'35
'37
'39
'41
'43
'45
'47
'49
'51
'53
'55
'57
'59
'61
'63
'65
'67
'69
'71
'73
'75
'77
'79
'81
'83
1984
'85

Figure B–5

Appendix C

AN ORDINANCE DECLARING AN EMERGENCY:
PROVIDING FOR THE LOCATION AND CONSTRUCTION OF A CHANNEL
TO CARRY ANTICIPATED FLOOD WATER:
PROVIDING FOR THE AUTHORITY OF THE CIVIL DEFENSE DIRECTOR
OF CASSIA COUNTY: AND
PROVIDING AN EFFECTIVE DATE

PREAMBLE

WHEREAS the Cassia County Commissioners recognize that the snow fall for the winter of 1983-1984 was exceedingly high and has caused the Lower Goose Creek Reservoir to potentially exceed its capacity; and

WHEREAS the Cassia County Commissioners recognize that flooding from the reservoir is imminent; that homes, property, and thousands of acres of crop land will be damaged or destroyed, and that human life is threatened; and

WHEREAS the Cassia County Commissioners recognize the need for the preservation of order, public health, safety, and minimizing property damage in Cassia County; and

WHEREAS the Cassia County Commissioners have sought the expertise and advice of independent agencies relative to alternative solutions to the imminent flooding. Advice was received from the United States Government acting through the Department of Agriculture Soil Conservation Service, the Department of the Army, Walla Walla District Corps of Engineers, and excavation contractors; and

WHEREAS the Cassia County Commissioners also sought and received advice from interested and potentially affected citizens of Cassia County; and

WHEREAS the independent agencies have recommended to the Cassia County Commissioners that to minimize potential damage to property and the risk to lives of the citizens of Cassia County there be a channel constructed to carry the flood water from the Lower Goose Creek Reservoir to the Snake River; and

WHEREAS the Cassia County Commissioners have relied upon the advice of the independent agencies, along with all other information available to the Commissioners at the time of the adoption of this ordinance, and have determined that the construction of a channel generally along the route shown on the attached Exhibit "A" would minimize damage to property and the risk to lives of citizens of Cassia County and preserve public order, health and safety; and

WHEREAS the Cassia County Commissioners declare this to be a great public calamity in the nature of an anticipated extraordinary flood; and

WHEREAS Article 12, Section 2 of the Constitution of Idaho authorizes the Commissioners to exercise authority under County police powers;

NOW THEREFORE BE IT ORDAINED BY THE BOARD OF COUNTY COMMISSIONERS OF CASSIA COUNTY, IDAHO;

Section 1: That a channel capable of carrying the anticipated flood water be immediately constructed from the Lower Goose Creek Reservoir to the Snake River across the lands as generally shown on the attached Exhibit "A".

Section 2: The Civil Defense Director of Cassia County is authorized to specify the precise location of the channel and to cause the channel to be constructed with the assistance of the Corps of Engineers and all other available aid. The Civil Defense Director is empowered to alter the general course shown on Exhibit "A" as necessitated by emergency, geographical and topographical conditions.

Section 3: The Commissioners do hereby declare that an emergency exists, which makes it imperative that this ordinance should become effective forthwith in order that the public health, welfare and safety might most effectively be provided for. Therefore this Ordinance is effective upon its adoption and posting as required by law.

ADOPTED this 16th day of May, 1984.

CASSIA COUNTY COMMISSIONERS
J. Weldon Beck—Chairman
Norman Dayley
John Adams

ATTEST:
Frank B. Kearns
County Clerk

Sources Consulted

Journals, notes, letters, reports, interviews with participants, and newspaper reports were the main sources of information. Kent Foster's field notes, Terry Bingham's daily log, Oakley Canal Company minutes, Donald J. Clark's minutes of LDS *church emergency meetings, and private journal entries were particularly valuable. Because most of the journal excerpts were read to the author, the original spelling and punctuation in these entries have not been preserved.*

Books and Articles.

Arrington, Leonard J. "Idaho and the Great Depression." *Idaho Yesterdays* 8 (Summer 1969): 2-8.

Arrington, Leonard J. "The Mormon Settlement of Cassia County, Idaho, 1873-1921." *Idaho Yesterdays* 13 (Summer 1979): 36-46.

Barber, Floyd R. and Dan W. Martin. *Idaho in the Pacific Northwest.* Caldwell, Idaho: Caxton Printers, 1966.

Daughters of the Utah Pioneers, Cassia County Company. *A Pause for Reflection.* Provo, Utah: J. Grant Stevenson, 1977.

Lovin, Hugh T. "A 'New West' Reclamation Tragedy: The Twin Falls-Oakley Project in Idaho, 1908-1931." *Arizona and the West* 20 (Spring 1978): 5-24.

Maughan, Ralph W. *Anatomy of the Snake River Plain: An Amateur's View.* Pocatello, Idaho: Idaho State University Press, 1992.

Pella History Committee. *Pella.. And Now There Are Two.* Burley, Idaho: Burley Reminder, 1977.

Piper, Arthur M. "Geology and Water Resourses of the Goose Creek Basin, Cassia County, Idaho." Bureau of Mines and Geology, Bulletin No. 6. Moscow, Idaho: University of Idaho, 1923.

The First One Hundred Years: Cassia-Oakley Idaho Stake, 1887-1987. Burley, Idaho: Burley Reminder, n.d.

Wells, Merle and Arthur Hart. *Idaho.* Windsor Publications, 1985.

Newspapers.

Burley Bulletin. (Burley, Idaho), 1913-1923; May 1, 1945.

Burley Herald-Bulletin. (Burley, Idaho), Golden Anniversary Edition Aug. 1955; Progress Edition Sept. 1958; June 17, 1963; May 22, 1967; July 16, 1969; Sept. 8, 1969; Aug. 1970.

Church News. (Salt Lake City, Utah: Church of Jesus Christ of Latter-day Saints), May 27, 1984; Nov. 28, 1987.

Engineering News-Record. (New York:McGraw-Hill), May 31, 1984.

Inside BLM, June 1984.

Intercom. (Walla Walla, Washington: U.S. Army Corps of Engineers), June 6, 1984; July 3, 1984; Oct. 10, 1984; Jan. 23, 1985; Mar. 29, 1985; April 1988.

Oakley Herald. (Oakley, Idaho), 1910-1935.

South Idaho Press. (Burley, Idaho), Jan 19, 1972; Feb. 28, 1974; Aug. 22-23, 1974; May 17, 1977; May 28, 1978; Mar. 25, 1982; Aug. 22, 1983; Jan. 5, 6, 11, 16, 1984; Feb. 20, 23, 1984; Mar. 4, 20, 1984; April 5, 18, 26, 30, 1984; May 1-31 (except 5, 12, 19, 26 when paper was not published), 1984; June 1, 3-8, 10, 12, 15, 1984; July 9, 11, 13, 1984; Aug. 1, 5, 7, 30, 1984; Sept. 4, 6, 26, 27, 1984; Oct. 8, 9, 11, 19, 1984; Nov. 8, 1984; Jan. 3, 16,

1985; Feb. 28, 1985; Mar. 20, 1985; Apr. 4, 1990; July 2, 1990; Oct. 29, 1990; Nov 12, 1990; Jan. 10, 1992; Mar. 8, 1992.

South Idaho Press, Minidoka County News, Farm Edition. (Burley, Idaho), March 1978, March 1984, June 1984, July 1984, March 1985, March 1988.

The Idaho Statesman. (Boise, Idaho), Jan. 4, 1984; May 4, 1984

The Times-News. (Twin Falls, Idaho), July 25, 1954; Apr. 28, 1984; May 3, 11, 16-21, 24; June 6, 1984; Jan. 28, 1985; July 2, 1990.

United States Government Publications

U.S. Army Corps of Engineers. Walla Walla District. "Flood Emergency Assistance." Walla Walla, Washington: April 1984.

U.S. Army Corps of Engineers. Walla Walla District. "Information Paper." Walla Walla, Washington: July 5, 1984.

U.S. Department of Agriculture. Soil Conservation Service. "Water Supply Outlook for Idaho." Boise, Idaho: Jan. 1, 1984; Feb. 1, 1984; Mar. 1, 1984; Apr. 1, 1984; May 1, 1984; June 1, 1984.

U.S. Department of Agriculture. Soil Conservation Service. "Snow Surveys and Water Supply Forecasting." Soil Conservation Service Information Bulletin 536. Washington, D.C., June, 1988.

U.S. Department of Interior. Bureau of Reclamation. *Minidoka Power Plant.* Washington D.C.: GPO, 1978.

U.S. Department of Interior. Bureau of Reclamation. *Minidoka Water Resource Project.*

Television Program.

"McNeil-Lehrer News Report." Directed and produced by Robert McNeil. PBS. KUTV, Salt Lake City. May 23, 1984.

Unpublished Sources.

Korsmo, Amund Marius. "The Twin Falls-Oakley Irrigation Project." Master's thesis, University of Illinois, 1913.

Local, State, and U.S. Government Files.

Cassia County Civil Defense. Flood file, which includes daily notes and video comments by Terry Bingham. Law Enforcement Building, Burley, Idaho.

Idaho Department of Water Resources. Oakley dam file, which includes reports of inspections of Oakley reservoir as well as miscellaneous papers relating to the flood. Department of Water Resources Office, Twin Falls, Idaho.

Idaho Department of Water Resources. 1984 flood file, which contains David Hollingshead's report in response to a Tort Claim. Department of Water Resources Office, Boise, Idaho.

University of Idaho Research and Extension Center. Daily records of stream flow for Goose Creek and Trapper Creek, and Oakley reservoir elevations from 1912-1986. Computer printout. Kimberly, Idaho.

U.S. Army Corps of Engineers. Public Relations Department. Correspondence with author, and miscellaneous papers.

U.S. Department of Agriculture. Soil Conservation Service. Flood file, which includes field notes by Kent Foster and miscellaneous papers. SCS Office, Burley, Idaho.

U.S. Department of Agriculture. Agricultural Stabilization and Conservation Service. Flood file. ASCS Office, Burley, Idaho.

U.S. Department of Commerce. Weather Forecast Recording. Regional weather data for 1983-1984. Weather Forecasting Office, Kimberly, Idaho.

U.S. Department of Interior. Bureau of Reclamation. Flood file, maps, and miscellaneous information on Goose and Trapper creeks. Minidoka Project Office, Burley, Idaho.

U.S. Department of Interior. Geological Survey. Files on Goose and Trapper creeks. Twin Falls, Idaho.

Minute Books of Organizations.

Burley City Council. Minutes of meetings, 1984. Burley City Hall, Burley, Idaho.
Cassia County Commissioners. Minutes of meetings, 1984. Cassia County Court House, Burley, Idaho.
Church of Jesus Christ of Latter-day Saints. Oakley, Idaho Stake. Minutes of stake, bi-regional, and landowners meetings pertaining to the flood; also correspondence and other miscellaneous papers.
Milner Low Lift Irrigation District Board of Directors. Minutes of special meeting, May 8, 1984, and agreement with Oakley Canal Company. Milner Low Lift Irrigation District Office, Milner Dam, Idaho.
Oakley Canal Company Board of Directors. Minutes of meetings, 1909-1926, 1934-36, 1983-1985; also miscellaneous papers concerning stream flow and water shares. Oakley Canal Company Office, Oakley, Idaho.
Oakley City Council. Minutes of meetings, 1984. Oakley City Office, Oakley, Idaho.
Oakley Highway Department Board of Directors. Minutes of special meeting May 8, 1984, and agreement with Oakley Canal Company. Photocopy given to author by Sharon Kimber, secretary.
Twin Falls Canal Company Board of Directors. Minutes of meetings, May 8, 1984; June 12, 1984; and Memorandum Agreement with Oakley Canal Company, May 9, 1984. Twin Falls Canal Company Office, Twin Falls, Idaho.

Private Journals and Papers.

Adams, John. Journal. Oakley, Idaho.
Bean, Delma. Journal. Burley, Idaho.
Bowers, Lila. Journal. Burley, Idaho.
Critchfield, Daren. Account of raising the spillway. Oakley, Idaho.
Gerratt, Marjorie. Journal. Burley, Idaho.
Gochnour, Cleo. Journal. Burley, Idaho.
Gochnour, Lane. Notes for speech given in church in 1985. Burley, Idaho.
Hardy, Karlene. Journal. Oakley, Idaho.
Harris, Randall. Journal. Burley, Idaho.
Hedberg, Kathleen. Journal. Burley, Idaho.
Henderson, Dorothy. Journal. Burley, Idaho.
Melling, Scott. Journal. Burley, Idaho.
Mitchell, Ellen. Journal. Oakley, Idaho.
Mitchell, Morris. Journal. Oakley, Idaho.
Mullen, Wayne. Journal. Oakley, Idaho.
Jenks, Nona. Journal and picture captions. Oakley, Idaho.
Randall, Julie. Journal. Burley, Idaho
Robins, Gloria. Journal. Oakley, Idaho.
Van Tassell, Carl. Journal. Oakley, Idaho.

Interviews

Adams, Douglas. 8 December 1991.
Adams, Brett. 21 February, 1992.
Adams, John. 28 February 1991.
Adams, Kelly. 21 February 1991.
Adams, Maxine. 3 February 1993.
Adams, Larry. 21 February 1991.
Adams, Leon. 23 January 1992.
Adams, Pat. 23 January 1992.
Allphin, Blaine. 8 January 1992.
Altom, Denver. 25 November 1990.
Anderson, Alberta. Telephone interview. 13 May 1992.
Baker, Leland. 8 December 1990.
Barrus, Alfred. 18 October 1990.

Barrus, Kathleen. 25 October 1990.
Beams, Roy. 28 November 1990.
Bean, Delma. 14 July 1991.
Bean, Roland. 25 January 1991; telephone, 1 May 1992.
Beck, Bruce. 7 May 1992.
Beck, David. 5 February 1992.
Beck, Dawn. 7 May 1992.
Beck, J. Weldon. 18 June 1990.
Beck, Susan. 5 February 1992.
Bedke, Karl "Buddy." Telephone interview. 5 February 1992.
Bingham, Terry. 4 June 1990, 28 January 1991, 23 April 1991, 17 June 1992.
Blacker, Lovel. 13 June 1990, 8 March 1993.
Blacker, Marvin. 13 June 1990.
Blauer, Norene. 28 July 1991.
Blauer, Willes. 28 July 1991.
Bowers, Bart. 13 March 1992.
Bowers, Lila. 13 March 1992.
Bright, James. 30 January 1991; telephone, 18 June 1992.
Burch, Lawrence. 4 April 1991.
Busch, Leo. Telephone interview. 5 January 1991.
Bywater, Stephen. 5 February 1991.
Campbell, Michael. 8 June 1992.
Clark, Donald. 26 June 1992.
Clark, Ed. 12 July 1990.
Clark, Eula. 27 June 1990.
Clark, John. 27 June 1990.
Clark, Thelma. 12 July 1990.
Corless, Earl. 26 October 1990, 29 October 1990, 24 March 1992.
Cranney, Bill. 12 July 1990.
Cranney, Douglas. 12 July 1990.
Cranney, Kim. 12 July 1990, 6 December 1990.
Cranney, Mike. 1 February 1991.
Critchfield, Daren. 5 December 1990.
Critchfield, Suzanne. 25 November 1990.
Crystal, Billy. 2 June 1990, 29 April 1992.
Curtis, Burdell. 21 October 1991.
Davidson, Ezra "Bud." Telephone interview. 6 November 1991.
Davis, Denny. 29 October 1990, 24 March 1992, 13 May 1992.
Dayley, Norman. 26 August 1992.
Dick, David. 24 June 1992.
Dunn, A. Kenneth. Telephone interview. 7 August 1991.
Easton, Joan. 18 June 1992.
Elquist, Dallan. 4 March 1991.
Elquist, Norma. 4 March 1991.
Evans, John, Sr. 20 March 1991.
Fairchild, Julie. 3 October 1991.
Foster, Kent. 7 November 1990.
Fronk, Kenneth. 14 November 1990.
Fuller, Ken. 16 October 1990.
Geary, Connie. 3 January 1992.
Geary, Tom. 3 January 1992.
Gerratt, Barbara. 25 February 1991.
Gerratt, D. Rex. 27 March 1990, 6 March 1992, 21 April 1992, 9 May 1992.
Gerratt, Larry. 25 February 1991.
Gerratt, Marjorie. 6 March 1992, 21 April 1992, 9 May 1992.

Gibbons, David. 30 October 1990.
Gochnour, Carla. 5 December 1991.
Gochnour, Cleo. 27 August 1991.
Gochnour, Jack. 27 August 1991.
Gochnour, Lane. 5 December 1991.
Gochnour, Sherry. 27 August 1991.
Gooch, Carl. 19 February 1991.
Gorringe, Loreen. 26 July 1991.
Handy, Ernest. 2 October 1991.
Hardy, Karlene. 30 April 1992.
Hardy, Randy. 30 April 1992.
Hardy, Richard. 22 April 1991.
Harmon, Lynn. 28 November 1990.
Harris, Edwin. 30 January 1991.
Harris, Marian. 30 January 1991.
Harris, Randall "Randy". 27 March 1993.
Hawker, Fred. 15 April 1991.
Hawker, Joan. 15 April 1991.
Henderson, Dorothy. 30 August 1990.
Higens, Jim. 10 July 1992.
Hill, Lida. 5 July 1990.
Hoffbuhr, Bud. 27 May 1991.
Hollingshead, David. September 1, 1992. Telephone interview. 16 May 1992.
Holmes, Curtis. 30 November 1992.
Holmes, Don. Telephone interview, 1 December 1992.
Holmes, Loren. 22 November 1991.
Hurst, Norman. Telephone interview. 14 November 1990.
Hurst, Timothy. 2 February 1991.
Jenks, Nona. 2 February 1991.
Jenks, Orvin. 2 February 1991.
Jensen, Glen. 29 June 1990.
Jensen, Marge. 14 July 1990.
Kearns, Frank. 5 June 1990.
Kimber, Ray. 26 January 1991.
Kimber, Sharon. 26 January 1991.
Klaas, Maurice. 28 November 1990.
Lattimer, Martha. 21 June 1990.
Layton, Rita "Skeet." 6 June 1990.
Lloyd, Delbert. 2 January 1992.
Lloyd, Max. 2 January 1992.
Luke, Eola. 12 November 1991.
Luke, Gordon. 12 November 1991.
McLaws, J. J. 22 June 1990, 13 February 1991.
McFarland, Amber. 30 January 1991.
Martin, John. 13 November 1991.
Merritt, Allen. 22 November 1991.
Messley, Inez. 14 May 1991.
Messley, Leman. 14 May 1991.
Mitchell, Morris. 31 May 1992.
Moncur, Gale. 30 January 1992.
Mullen, Elda. 27 April 1992.
Mullen, Wayne. 27 April 1992.
Nelson, Newel. 17 July 1990.
Parsons, William. 17 April 1991.
Pearson, Steven. 16 March 1993.

Petersen, Eileen. 16 July 1990.
Petersen, Walter. 12 July 1990.
Pickett, Ennis. 15 November 1990.
Pinnock, Hugh. Telephone interview. 15 August 1990.
Poulton, Boyd. 20 February 1991.
Poulton, Kathy. 27 July 1992.
Poulton, Kelly. 3 February 1993.
Poulton, Mike. 27 July 1992.
Pressley, LaVonne. Telephone interview. 11 July 1991.
Randall, Julie. 2 November 1991.
Randall, Richard. 12 June 1990.
Randall, Rosanne. 12 June 1990, 15 February 1993.
Rasmussen, Evan. 7 November 1990.
Rice, William. 16 October 1990.
Robins, Gloria. 15 January 1992.
Robins, Val. 15 January 1992.
Robinson, Jay "Bud." 12 February 1991.
Robinson, Reine. 12 February 1991.
Roskelley, Darrell. 16 February 1991.
Sagers, Robert. 6 June 1991.
Schloss, Margaret. 30 September 1990.
Scholer, Burke. 7 November 1990; telephone interview 26 April 1992.
Searle, Kent. 19 April 1991.
Shadduck, Chuck. 10 April 1991.
Siepert, Milo. Telephone interview. 14 January 1992.
Smith, Larry. Telephone interview. 4 January 1992.
Spurgeon, Olene. 5 September 1991.
Stanger, Clem. 13 May 1992.
Stevenson, John "Bert." 15 January 1992.
Stock, Merlin. 12 October 1990.
Straubhaar, Ernest. 25 March 1991.
Sullivan, Evelyn. 28 June 1990.
Sullivan, Grant. 28 June 1990.
Tate, Tom. Telephone interview. 15 June 1990.
Taylor, Carolyn. 8 January 1991.
Taylor, Cloyd. 22 February 1991.
Thomas, Neal. 3 October 1991.
Tilley, Melvin. Telephone interview. 18 June 1992.
Van Tassell, Carl. 14 February 1992; 27 January 1993.
Van Tassell, Elda. 14 February 1992; 27 January 1993.
Wake, Klyda. 23 June 1990.
Warr, Viva. 6 June 1990.
Watterson, Paula. 15 May 1993.
Wayment, Dal. Telephone interview. 7 January 1991.
West, Harold. Telephone interview. 8 January 1991, 25 April 1992.
Whiteley, Winslow. 23 June 1990.
Wickel, Wynell. Telephone interview. 29 February 1992.
Williams, Morgan. 7 November 1990.
Woodhouse, J. Allen "Duff." 13 November 1991.
Woolstenhulme, Alfred "Tex." 5 February 1992.
Wyatt, Grant. 14 January 1991; telephone interview 29 August 1991.